JOSEPH PRIESTLEY
A CELEBRATION OF HIS LIFE AND LEGACY

First published in 2007 on behalf of The Priestley Society
by Scotforth Books,
Carnegie House,
Chatsworth Road,
Lancaster LA1 4SL,
England
Tel: +44(0)1524 840111
Fax: +44(0)1524 840222
www.scotforthbooks.com

British Library cataloguing-in-Publication data
A catalogue record for this book is available from the British Library

ISBN 1-904244-44-0
EAN 978-1-904244-44-8

Front cover image of Joseph Priestley, known as the 'Leeds Portrait', is reproduced here courtesy of the Royal Society of London

Joseph Priestley
A Celebration of His Life and Legacy

Bicentennial Commemorative Collection

Edited by
James S. Birch and Joe Lee

Published by the Priestley Society

2.

Contents

List of Subscribers

The Priestley Society wishes to acknowledge, with many thanks and appreciation, the sponsorship of this publication by BOC. In particular the Society wishes to thank Drs Paul Page and Patrick Pearson of BOC Corporate Development for their support of the Priestley Society Symposium at the University of Huddersfield and for their constant encouragement in the preparation of this book.

BOC
Dr Williams's Trust and Library
Friends of Joseph Priestley House, Northumberland, Pennsylvania
Jane Gummer and Govenors of Batley Grammar School
The Leeds Library
Old Batelians (Old boys of Batley Grammar School)
The Rotary Club of Heckmondwike and District
The Royal Society of London
Woodkirk High Specialist Science School

Gary D Anderson	Grandmaster Alchemist, Alpha Chi Sigma, USA
Dorothy Archer	National Unitarian Fellowship
David Asquith	Heckmondwike Rotary Club
Bill and Jenny Bailey	Castleford, and Priestley Society
Douglas Baird	Toronto East General Hospital, Canada
Stephen H Bartle	Priestley Historian and Author, Birstall
Andrea Bashore	Joseph Priestley House, Northumberland, Pennsylvania
Heather Bear	Dickinson College, USA
James S Birch	Priestley Society
Lesley Birch	Priestley Society
Ray Birks	Priestley Society
Ronald and Frances Blatchley	Friends of Joseph Priestley House Museum, USA
Helen Braithwaite	Priestley Society
Malcolm J and Hazel Byard	Priestley Society
Dee Ann Casteel	Bucknell University, and Unitarian Congregation of the Susquehanna Valley, USA
Kath Conway	Priestley Society
R J S Crossley	University of York
Malcolm Dick	Joseph Priestley and Birmingham Project, and University of Birmingham
Jennifer Doran	Dickinson College, USA
Alan T Dronsfield	Historical Group, Royal Society of Chemistry

Roger Egolf	American Chemical Society, Division for the History of Chemistry
Peter Excell	University of Bradford
John and Claire E Fairhurst	Fairhurst Mouldings Ltd
Paul Goulden	History of Anaesthesia Society
Peter Haley	Rotary Club of Headingley
Michael and Barbara Hanson	Heckmondwike Rotary Club
Edward and Margaret Hill	Priestley Society
Eric J and Kath Hustwick	Priestley Society
Sandra Knapp	Natural History Museum
David Knight	History and Philosophy of Science, University of Durham
Alex Lee	Priestley Society
Joe Lee	Priestley Society
Edward Lyon	Priestley Society
John F Maguire	United States Air Force Research Laboratory
Stuart Martin	Priestley Society
Colin Moore	Menston, Ilkley
Alan Naylor	Priestley Society
Paul Neaves	Star Centre, Keighley College
Folu Ojutalayo	Huron University
Roy Olofson	Pennsylvania State University
Ann Peart	Unitarian College, Manchester
Suzanne Margaret Priestley	Priestley Society
Margaret M Rhodes	Bradford
David Richardson	Hanover, New Hampshire
Norman Rose	Priestley Scholar, University of Birmingham
John Rowlinson	Physical and Theoretical Chemistry Laboratory, Oxford University
Charles Soderlund	Priestley Society
Derek Stephenson	Heckmondwike Rotary Club
Kate Taylor	General Assembly of Unitarian and Free Christian Churches
Priestley Toulmin	Alexandria, Virginia,
Ian and Hilary Wainwright	Birstall
Margaret E Ware	Great-Great-Great-Great-Great Niece of Joseph Priestley
Yoshiaki Watanabe	Priestley Society
Ruth Watts	History of Education, University of Birmingham
Randy Wayne	Priestley Society
Les Woodcock	Priestley Society
Chris Wooller	Robertown, West Yorkshire
Amos J Wright	Birmingham, Alabama

Foreword

St. Matthew, Chapter 13, verse 57: And Jesus sayeth unto them
"A Prophet is not without honour, save in his own country and in his own town"

On the evening of March 13th 2003 a small group of well-wishers met in the Old Hall, Heckmondwike, West Yorkshire, Joseph Priestley's boyhood home. The topic of discussion was how we might celebrate the bicentenary of the life and legacy of Priestley (Born 13th March 1733 in Birstall; died 6th February 1804 Northumberland Pennsylvania USA) It was resolved that the forthcoming 2004 Bicentennial would not pass unheralded, at least in West Yorkshire. This Commemorative Collection represents the culmination of those early deliberations, three years on, and summarises in a single volume, the recorded contributions of many Priestley scholars, exhibitors and distinguished lecturers from the UK and North America.

Events began at 10.30 on the morning of Friday 6th February 2004 with a walk from the Priestley birthplace at Fieldhead to the Old Hall, now the home of the Priestley Society. The following day, celebrations continued with a colloquium of invited speakers and scholars at Oakwell Hall in Birstall. The same day, the Priestley Society held its Inaugural General Meeting and became an official body with a constitution and a mission statement. One objective of the Society's mission is to correct the myths that Priestley was some kind of unpatriotic rebel, and to reverse the institutionalised neglect of Britain's immensely rich Priestley heritage.

One only needs to examine BBC programmes to see the need for a Priestley Society. On April 7th 1994, the 200th Anniversary of the Priestleys' leaving for America, Priestley supporter, Bernard Howgate, wrote a letter to the Batley News. In that letter, he complained about the BBC. The title of his article was *Will Britain ever Honour this Man?*. A noteworthy citation in Mr. Howgate's letter is a BBC programme Horizon, broadcast on October 13th 1983 in which Dr Priestley was given faint praise for one scientific discovery, but denounced as a rebellious 'red', a 'left-wing' rebel. Although these terms didn't exist in Priestley's day, nothing could be further from the truth. He spent his whole life campaigning for civil, political, and religious liberty, and was indeed fearful of the corrupting role of Government in religion and education. He campaigned relentlessly in search of truth, and would have been opposed to any broadcasting organisation being used as an instrument of government.

Recently the BBC made popular capital by broadcasting a series of programmes to find *The Greatest Briton Ever*. Starting with a hundred names, programme by programme the BBC reduced the field to ten grand finalists included two scientists (Newton and Darwin), an engineer (Brunel), an explorer (Cook), a playwright (Shakespeare) a pop-singer (Lennon), an admiral (Nelson), a politician (Churchill) and royalty (Queen Elizabeth I and Princess Diana). Sir Winston Churchill won the *BBC's Greatest Briton Ever* nomination in the Grand Final screened on Sunday 2 February 2003.

How could the name of Joseph Priestley possibly not even be included in the BBC's starting line-up of all time British greats? These proceedings of the Bicentennial Year are now part of the testimony that the legacy of Joseph Priestley, in any single one of the areas of science, education, religion and politics, far outweighs the narrow achievements of any individual listed in the BBC's grand finalists, let alone the 100 starters from which he was excluded. Indeed, if one searched worldwide for the greatest achiever of mankind, it would be hard to find a stronger candidate than Joseph Priestley. He should be up there with the great world reformers who have contributed to civilization, from Socrates to Ghandi.

Part of the explanation undoubtedly rests in the old adage "Don't mix politics with religion". Why not, one might reasonably ask? Politics and religion are indeed inextricably linked and, moreover, the connections have bedevilled the Government of Britain for at least a thousand years and continue to do so. It is also often advised that one should not mix religion with science. Again, we might ask, why not? Perhaps it's because truth, in the form of natural knowledge, rests with the scientists, but power and influence remains with the religious bigots.

Priestley mixed religion, politics and science in a life of rich academic bequests. In seeking out the truth, he observed that the religious powers that be, within a corrupt 'Church and Monarchy' were suppressing advancements in science, education and civil liberty in the maintenance of their privileged power base for which there was neither legitimate constitutional nor rational basis. When there was no science there was no conflict. With the advent of science, the need for religion is debatable; nevertheless, religion remains an instrument of suppression and control in every country in the world today.

Priestley's earliest claim to greatness is in the field of education. As a young graduate from the dissenter's Academy at Daventry, he opened a private school in Nantwich, where the education was so enlightened that it began to attract attention over a wide area. After three years he was recruited as a tutor to Warrington Academy, which emerged as the leading nonconformist college in England. Over a period of six years, Priestley so modernised and improved education there, that Warrington became the finest centre of educational excellence in the known world, outstripping even those theological university establishments of Oxford and Cambridge. The globes that Priestley designed for teaching geography and astronomy, terrestrial and celestial respectively, are now in Harris Manchester College Oxford, and described here in these pages. In his postgraduate studies, Britain's only Priestley PhD scholar, Dr Norman Rose, concludes that Priestley greatly surpassed all other authors in the English speaking world combined, and that his achievements on education alone would qualify him for the BBC's Greatest Briton Ever.

Priestley's discovery of the element-of-life oxygen is legendary. What is not well known, however, except to historians of science, is that he also made fundamental experimental discoveries and inventions in many other areas of chemistry and related sciences. Besides oxygen, he discovered and characterized nine other gases, including the oxides of nitrogen and carbon, and ammonia. It has often been stated that his discovery of oxygen, and later nitrous oxide, the first anaesthetic, was a landmark in medicine. Priestley is recognised as a founder of modern medicine; he discovered that dark venous blood turns bright red in the presence of oxygen and that the process is reversible, and went on to explain the role of the blood in oxygen transport and even proposed that the body's use of oxygen and the oxidation of metals were similar phenomena. Priestley's curiosity of the new element oxygen led to his discovery of the process of photosynthesis and the production of oxygen by plant life. In 1768, Priestley deduced and published Coulomb's law of electrostatic forces 20 years before Coulomb; he discovered the electrical conductivity of charcoal, and he established the connection between electricity and chemical reactions when he decomposed ammonia with electricity, 70 years before Faraday.

If Newton and Darwin were Britain's founding fathers of Physics and Biology, respectively, Joseph Priestley was surely Britain's founding father of Chemistry. On the scientific criterion alone, why, we might ask the Royal Mint, is not Joseph Priestley up there as a great Briton alongside Newton (£1 note), Darwin (£10 note) and Faraday (£20 note)? In Appendix III we see that Britain's founding father of Chemistry was indeed commemorated on Britain's coinage, two halfpenny tokens. The first shows Priestley hanging from a gibbet for his alleged support for a British-style French revolution, and the second shows him as the head of a snake, responsible for attempting to burn down the House of Commons. Both implied allegations were, of course, without foundation.

In religion, Priestley played a key role in recognising and publicising that the basis of religious discrimination is power and control of peoples' lives, and exaltation of religious leaders. Priestley recognised that the degeneration of religions was more to do with perpetuation of power and corruption of principles, than theology. His reconciliation of different religions, advocacy of the separation of Church and State, call for the abolition of religious dogma, and foundation of Unitarianism, were major advancements in the history of religion. The First Unitarian church of Philadelphia originated in a series of sermons Priestley gave in that city and marked the beginning of Unitarianism in North America. Priestley's contribution to religion would also qualify his candidacy for one of the greatest Britons.

Yet, it is the political legacy of Joseph Priestley that puts him up there above Churchill; it has been of immense consequence in the USA, the country we must thank for victory in two world wars of the 20th Century. In 1768 Priestley published his *Essay on the First Principles of Government and on the Nature of Civil Political and Religious Liberty*. A year later, in 1769, he wrote *On the present state of Civil and Religious Liberty in Great Britain and her Colonies*. In the preface to the first Essay, he lamented that there was no nation in the world whose government satisfied the principles he advocated. Just eight years later came the American Declaration of Independence, and the nation of states, with the civil, political and religious liberties that Priestley had so precisely described, became the reality. His political ideas on the importance of separation of Church and State, civil liberty, and the concept of a Supreme Court appointed by elected representatives, for example, were accepted by the Founding Fathers of the United States and enshrined into the US Constitution and Bill of Rights.

In the following pages, the Priestley Society brings a new dimension of truthfulness, in the spirit of Priestley himself, to the recognition and celebration of his life and the promotion of his ideals. The Priestley heritage trail begins, in West Yorkshire. This "Life and Legacy" collection hopefully, will be seen as a landmark in the beginning of the re-establishment and reconciliation of Joseph Priestley, as a British patriot and 'Citizen of the World' (see APPENDIX III). 2004 marks the beginning of a campaign by the Priestley Society to educate the people of Great Britain, and the wider international community.

Henceforth, March 13th has been designated Priestley Day in West Yorkshire. July 4th will also mark an annual celebration, not just in the United States. For the first time in 2004, as described in these pages, we saw American Independence celebrated in the United Kingdom. We hope also that the first Priestley Society July 4th Annual Lecture, presented by Priestley's descendent Dr. Priestley Toulmin, and published here, will become an annual thank-you to Joseph Priestley for his great achievements and for his unique contribution to the civilisation of humanity.

Les Woodcock: Chairman of the Priestley Society, July 2005.

Postage stamps depicting Joseph Priestley (*centre*) and two of his collaborators, Thomas Jefferson (*left*) and Benjamin Franklin (*right*)

Heritage Walk

Priestley Society members, local historical groups, well-wishers, and visitors from far and wide, including 4 from Pennsylvania USA, and approximately 30 children from Fieldhead Junior School, assembled at Priestley's birthplace in Fieldhead at 10am on Friday, February 6th, 2004 to take part in a Heritage Walk, in commemoration of the life and legacy of Joseph Priestley.

The map on the next page shows the route of the walk; from the Priestley birthplace at Fieldhead to the Old Hall Heckmondwike, via the Priestley Statue in Birstall Market Place and the Priestley ancestral family grave in the grounds of St. Peter's Parish Church of Birstall.

The Priestley Society wishes to acknowledge the following supporters of the Heritage Walk: Rev Patrick Senior of Brownhill Church for organising the events on the Market Place; Melanie Prideaux of Yorkshire Unitarians for leading the Unitarian Service; Rev. Paul J. Knight for provision of the sound system; Rev. Keith Pratt and Christine Edmunds of the Methodist Church for the provision of refreshments; West Yorkshire Police for their duty as escorts; and the St. John's Ambulance service for their presence throughout the walk.

Priestley's Birthplace

A small plaque, barely readable from the roadside, has been fixed above the front entrance to the second of the three cottages. It just says "*Joseph Priestley Discoverer of Oxygen Born on this Site AD1733*". This terse inscription on the birthplace of one of the world's greatest scientists and political reformers seems

Joseph Priestley's Birthplace. Birstall England 1738

Artist's impression of Priestley's birthplace.

PAINTING BY STEPHEN BARLOW

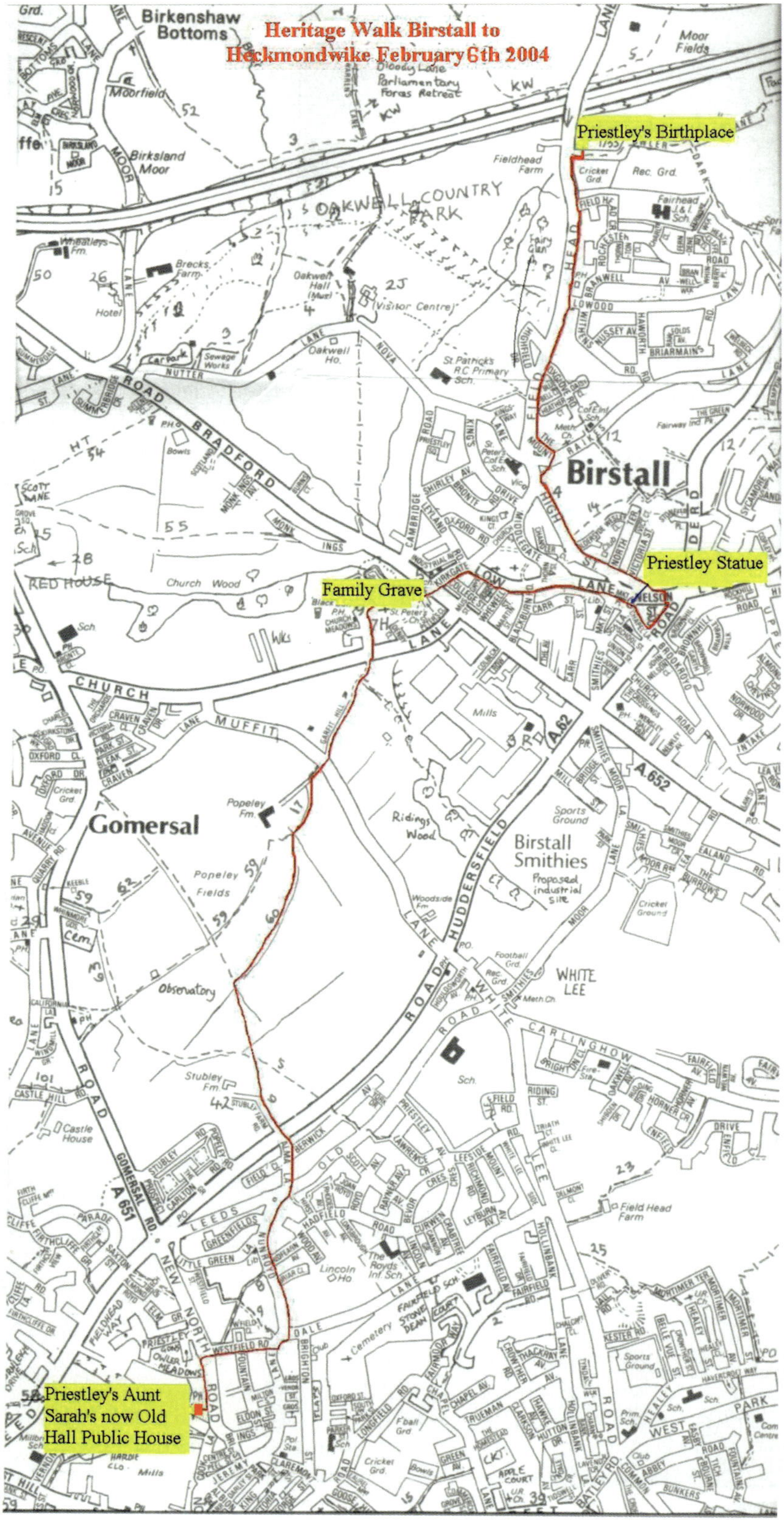

Heritage Walk
February 6th 2004

Right: Group of school children (with Roy Olofson) on the walk.
PHOTO COURTESY M.A. BASHORE

Below: Plaque at the site of Priestley's birth

to epitomise the institutionalised neglect, by both national and local authorities, of this historic heritage landmark.

Priestley's father, Jonas Priestley, processed woollen pieces in the workshop alongside his home; he employed approximately 10 people.

Fieldhead to Birstall

The first stop was at a farm at the bottom of Raikes Lane; this was a working farm in the 18th Century, but is now a place of residence. The walk continued down High Street with a short pause at Birstall Hall, further down the High Street, past the old blacksmith's shop and onto Huddersfield Road. The next call was at John Nelson's study; (John Nelson was a leading supporter of John Wesley, the founder of the Wesleyan church movement). From here the walk progressed to the statue of Priestley in Birstall Market Place.

At the Priestley statue in Birstall Market Place: Jim Birch with school children.
PHOTO COURTESY M.A. BASHORE

Market Place and Unitarian Service

A class of 30 children from Birstall Primary School joined in at the Market Place where Melanie Prideaux, a lay preacher with the Unitarian Church, conducted a service detailed below. At this point as if to order, the sun came out and shone brightly for the rest of the walk.

After the service, Birstall Methodist church members provided refreshments; this included for the children a fizzy drink. Joseph Priestley invented the process of carbonation and is therefore the founder of the "fizzy drinks industry".

The walk continued, without the children, through Birstall and down to St Peter's churchyard to look at the Priestley family grave; the gravestone had been restored with funds from the Priestley Society. We also looked at the grave of Helen Nussey, a close friend of the Bronte sisters. At one side of the graveyard is John Nelson's grave.

At this point our guide, Malcolm Haigh, explained the importance of Birstall in Priestley's time. St Peter's church was the main church in the area and people came from miles around to worship there. Birstall was also exceptionally well connected by the road network; anyone going north to south or east to west had to pass through the town. Also, the presence of a stream enabled the processing and dyeing of wool.

From the church the walkers climbed Garfitt Hill, up to Popeley and Stubley farms, working farms in the 18th century. From here we continued down to the Old Hall in Heckmondwike; this was Priestley's Aunt Sarah's home where Priestley lived for many years (see paintings on page 8).

The Service in Birstall Market Place

Mel Prideaux

Led by Mel Prideaux, Lay Leader, Westgate Unitarian Chapel, Wakefield.

- Words of welcome and prayer.
- Brief address.
- Prayer.

For all, like Joseph Priestley, who have pushed back the boundaries of human knowledge in the search for truth

All: We give thanks

For all who have striven for human freedom and human dignity through the centuries

All: We give thanks

For those who named and pioneered the rights of conscience and self-determination

All: We give thanks

For those who asserted the freedom of mind and intellect and who challenged ignorance

All: We give thanks

For those who struggled and suffered to win us democracy, free speech, religious liberty and equality before the law

All: We give thanks

For all those who fought against slavery, tyranny and oppression

All: We give thanks

As we celebrate the achievements of Joseph Priestley, let us also accept our responsibility to enshrine his courage, openness and commitment to truth in our hearts and make it real in our lives. And may we accept our responsibility to live truly as members of one human family

All: As we give thanks, so we accept these responsibilities. Amen

We stand in silence for a short while, and then join in saying the words of the Prayer of Jesus:

Our Father, who art in heaven, hallowed be thy name.

Thy kingdom come, thy will be done, on earth as it is in heaven.

Give us this day our daily bread and forgive us our trespasses, as we forgive those who trespass against us.

Lead us not into temptation, but deliver us from evil, for thine is the kingdom the power and the glory, forever and ever.

Priestley Family Grave, before *(above)* and after *(below)* repairs

Amen

Priestley Family Grave: St Peter's Churchyard

The family grave is in St Peter's Churchyard Birstall. This was in a bad state of repair and no funds were available from the Church, so the Priestley Society raised £352.50 to have it refurbished, just in time for the celebratory walk on 6th February 2004 to mark 200 years since Joseph's death.

Grave Inscription:

Here Lyeth interred
Body of Sara wife
of joseph Priestley of Bir
stal feild Head who dyed
29th of December 1728
Aged 2?8 Years
On North side Lyethy
Body of joseph Son of jos
eph Priestley who dyed 16th
of August 1713 Aged 25 years
Ony South Side Lyethy
Body of Phebe Daughter
of john Priestley who dyed 4th
of February 17?0 Aged 1
Year And 6 month
Here alfo is Interred
the Body of Joseph
Prieftley of Birftal feild
head who departed this
Life the 2nd day of Auguft
in the 85th Year of his
Age Anno Dom 1745

Joseph, son of Joseph, is recorded here as having died on 16th August 1713 aged 25 years.

Not 6th November 1706 as shown on the family tree in Appendix IV.

Joseph (the senior) is recorded here as having died on 2nd August.

Not 5th August as shown on the family tree in Appendix IV.

Plaque on the wall of the Old Hall

Inn sign, Old Hall, Heckmondwicke

Old Hall Public House, Heckmondwike

Within the Old Hall, on the day of the walk, various exhibits were on display in the upstairs function room, which has now been named the Priestley Room. Amongst the exhibits were prints of the 10 paintings by Stephen Barlow, and a poster by Roy Olofson on the coins and medallions that commemorate the life of Joseph Priestley. Some of the medallions on the poster are reproduced in APPENDIX III.

Joseph Priestley at the Old Hall, Heckmondwike. 1748.

The Old Hall, Heckmondwike.

Two paintings of the Old Hall by Stephen Barlow, the first giving an impression of the building in Priestley's time and the second giving a present day scene

SECTION 2

Bicentennial Banquet

JOSEPH PRIESTLEY, CITIZEN OF THE WORLD – A CELEBRATION OF HIS LIFE

Opening Address by Dr. Stephen Bartle,
Vice-Chairman Priestley Society, Master of Ceremonies

We have gathered here to recognise the achievements of Dr. Joseph Priestley and to celebrate his life on the occasion of the bicentennial of his death.

From a somewhat difficult childhood characterised by the loss of his mother, and a degree of recurrent ill-health, this remarkable man emerged possessed of a powerful disciplined and perceptive mind, accompanied by a drive and enthusiasm to produce a prodigious if not unique harvest of discovery, invention and achievement in so many diverse fields of scholarship, that only those specialists in specific disciplines can begin to appreciate the full range of his undoubted genius.

What is more, and perhaps without parallel is his enormous devotion to informing others of the basic foundations of science in its many manifestations, providing guidance for the beginner in a particular scientific discipline. Priestley would discuss it, reflect on it, establish its nature and write it up, devising a programme of research based on observation and experiment. For Priestley was not only a scientist but an educator with an instinctive drive to inform, inspire and stimulate his readers and students. He aimed to establish a disciplined and meticulously conducted scientific method in all his experiments and theories: leavened with a challenging enthusiasm. It would be hard to find a famous scientist of Priestley's day to rival the quality and quantity of his publications ... Above all, he wished to encourage others to publish and circulate their discoveries.

The guiding principle of Priestley's life was peaceful intercourse between the nations in commerce and in scientific knowledge. Priestley is totally innocent of seeking any form of primacy in the intellectual and scientific world of his day. Nor at that time, was there any evidence of nationalism in terms of discovery; rather such achievements were seen as the common property of humanity in general and open to a sort of corporate enterprise (non-profit-making) to extend knowledge as an international endeavour to promote human well–being through the advancement of Science. This is reflected in the promotion of scientific advances under the French Revolution and Empire which involved the dissemination of information by government funded research institutes.

It is too easy for us to see Priestley and his age through 21st century eyes, which are clouded with a tendency to selfish materialist considerations, and also with ambitions of a purely personal and national nature. To treat the contentious subjects of the discovery of Oxygen, or who should qualify for the title of "Father of Modern Chemistry" seem to lose their significance when we remind ourselves that it never occurred to Priestley to seek anything of the sort. Such gestures spring from the desire of Priestley admirers to recognise the immense debt that we owe to him.

If Priestley had kept quiet about his political convictions and concentrated on his scientific, theological and educational work then he might have continued to flourish in the scientific circles of Britain. However, the very basis of the spirit of Priestley's energy in science was his Christian conviction … "To labour and not to seek for any reward"… and his insistence that "all people should have an expectation of happiness" *i.e.* through democratic process. This sealed his fate in 1791 and his life's work *i.e.* his laboratory, his library and his correspondence were destroyed and he barely escaped with his life.

Let us therefore celebrate the achievements of this great man as an outstanding leader, completely selfless in his motives to promote the happiness of mankind and to support in every way possible his fellow scientists of all nations.

This was followed by an address by the Mayor of Kirklees and a short response by Dr. Bartle

Proclamation from Pennsylvania

A Proclamation by Andrea Bashore Director of
Joseph Priestley House Museum Pennsylvania
On behalf of
Pennsylvania Historical and Museum Commission Resolution

Pennsylvania Historical and Museum Commission

Resolution

WHEREAS, Joseph Priestley, noted British theologian, natural philosopher, educator and political theorist was born on March 13, 1733 at Fieldhead near Leeds, West Yorkshire, England, emigrated to the United States of America on April 8, 1794 and resided at Northumberland, Pennsylvania where he died on February 6, 1804; and

WHEREAS, the year 2004 marks the 200th anniversary of Joseph Priestley's passing; and

WHEREAS, The Priestley Society of Birstall and Heckmondwike, West Yorkshire, England wishes to commemorate the life of Joseph Priestley during this 200th anniversary year with a special weekend of historical activities, meetings, exhibitions and colloquium highlighting Joseph Priestley's achievements; and

WHEREAS, The Priestley Society wishes to establish a formal contact with the Pennsylvania Historical and Museum Commission, the Joseph Priestley House, and the Borough of Northumberland, Pennsylvania; and

WHEREAS, the Pennsylvania Historical and Museum Commission administers the Joseph Priestley House in Northumberland as a memorial to Joseph Priestley, a figure of great historical significance in the inquiries of theology, chemistry, education and political thought;

NOW, THEREFORE, BE IT RESOLVED, that the Pennsylvania Historical and Museum Commission sends greetings and congratulations to the Priestley Society of Birstall and Heckmondwike on the occasion of this special program commemorating the life and contributions of Joseph Priestley on February 6, 7, and 8, 2004 in West Yorkshire, England, and acknowledge and applaud the Society's efforts to forge an international relationship with shared goals of preserving and interpreting the legacy of Joseph Priestley for generations yet to come.

I hereby certify that the above is a true and correct copy of the Resolution adopted by the Pennsylvania Historical and Museum Commission at its November 19, 2003 meeting.

Wayne S. Spilove
Chairman

John Wesley
Interim Executive Director

SECTION 3

Oakwell Hall Colloquium

OAKWELL HALL

The Hall stands in a country park of approximately 110 acres and was built in its present form around 1583 by John Batt. There was a house on this site from 1350. John Batt (1535–1607) inherited the estate in 1572; his initials and the date can be seen over the porch doorway leading to the Hall. Parts of the building were added at a later date.

John Batt's grandson, John Batt II, probably took part in the English Civil War, at the battle of Adwalton Moor (30th June 1643), about 1 mile from the Hall and was a victory for the Royalists.

Over the years various occupants rented the Hall, including Thomas Clapton. He shared the Hall with his cousin Sarah Nussey; Sarah was aunt to Helen Nussey, friend of Charlotte Bronte. Charlotte described Oakwell as 'Fieldhead', home to Shirley Keeldar, in her novel 'Shirley' published in 1849.

In 1926, wealthy Americans tried to buy the Hall with the intention of taking it, piece by piece, to the United States of America. An appeal was started to raise funds to buy the Hall so that it could remain in Birstall. Fortunately, Sir Norman Rae and Mr John Sharman bought and restored the Hall, and then gave it to Birstall Urban District Council.

The Hall was opened as a public Museum and is now the property of Kirklees Metropolitan Council.

Oakwell Hall

PRESENTATION TO THE PRIESTLEY SOCIETY

During the Oakwell Hall meeting, Professor Roy Olofson, on behalf of Pennsylvania State University and the Historical Division of the American Chemical Society, presented the Chairman of the Priestley Society, Professor Les Woodcock, with a medallion struck by the Franklin Mint of Philadelphia in 1974 to commemorate the 200th Anniversary of the Discovery of Oxygen.

The Chairman, on behalf of the Membership of the Priestley Society, duly thanked Professor Olofson for all his support and agreed, on behalf of the Priestley Society and all its members, to keep the medal, as a symbol of international recognition, in perpetuity for so long as the Priestley Society shall be in existence.

The Presentation medal

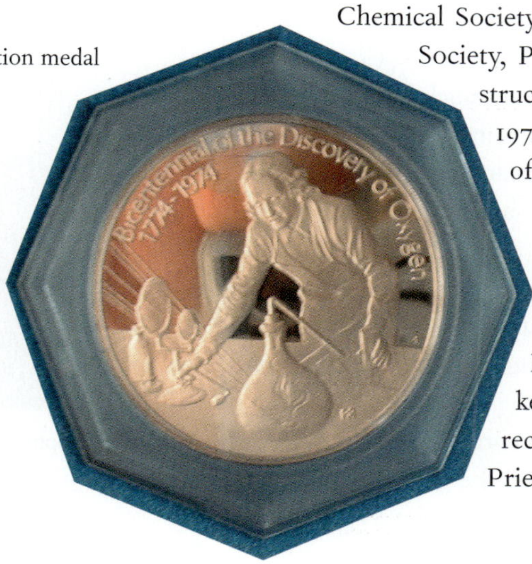

The presentation: Roy Olofson (*left*) and Les Woodcock (*right*)

THE VINDICATION OF JOSEPH PRIESTLEY, THE GREATEST BRITON EVER!

Stephen Bartle
Priestley Historian, University of Leeds (retired)

Stephen Bartle

In the year of Joseph Priestley's Bicentennial, much has been done to make it a memorable occasion and many new insights into the genius of this polymath of polymaths have come to light, such as his anticipation of the Coulomb Law of electrostatics and his profound influence on the process of education.

However, in one fundamental respect in all the articles and tributes to him, we have failed Priestley dismally. The question we must ask ourselves is: why should this successful man of science and so many other disciplines jeopardise his achievements by taking on the establishment in the matter of parliamentary reform?

William Wilberforce, early in his parliamentary career, introduced a bill for the reform of the penal code. When it failed, he gave it up and turned to the abolition of slavery (which was supported by Priestley).

Why then should Priestley pursue reform? The answer is because it offended his Christian conscience and sense of injustice. He held the view that governments should attend to the happiness of ordinary people; a matter which never concerned those who participated in the management of affairs.

So, Priestley was willing to press his energies to exposing the root cause of suffering endured by the common people who had no say in their future.

As a consequence, he was recognised as the leader of the movement towards Parliamentary reform by the King and the government. This singled him out for special attention. There is considerable evidence that Priestley was made the victim of a cold-blooded deliberate plot to mislead the public by falsely displaying him, not as the enlightened reformer, but as a rabid republican, hell-set on the execution of the King and the establishment of a republican government. This was conducted with icy skill and ruthless planning. A violently seditious handbill was printed and distributed around the Midlands and a false rumour was widely spread throughout the region to the effect that Priestley ('Gunpowder Joe') had toasted the King's head on a plate! This, on July 14 1791, preceded the second anniversary of the Storming of the Bastille.

This plot was not confined to the provinces, for James Gillray in London produced an elaborate cartoon – see opposite – which appeared simultaneously entitled "The Hopes of the Party Prior to July 14 1791". This shows the Whigs beheading George III at the block, with Charles Edward Fox – the leader of the Whigs – thinly disguised in a mask acting as executioner and referring to Charles I, and the Prince of Wales holding the King's head while Priestley offers comfort to the King: "My dear brother, we must all die once ... in fact a man ought to be glad if he can serve his country by bringing about a glorious revolution ...". For such service Gillray was to receive a pension of £200 per year.

The Gillray cartoon

All these infamous lies were dispensed to English people when the French king was about to accept constitutional government based on manhood suffrage and abandon the old absolutist monarchy.

The Priestley riots which were the consequence of the whipping up of 'popular fury', as they were described in the pro-government press, resulted in the destruction of the Unitarian meeting houses – Priestley was a Unitarian – and of Priestley's home and life's work.

It was no casual riot but a cold, deliberate act of wanton intimidation. Priestley was convinced that the authorities were involved in the plot and the Gillray cartoon suggests that the government lay behind its orchestration.

The poet William Wordsworth, in a reasoned account, supported Priestley's suspicions that local magistrates appointed by the Home Secretary and, working on behalf of the government, had a major part in this misinformation which inflamed the mob against Priestley. It has been established that the British Secret Service at that time had two functions: information gathering and, in this situation, the dissemination of misinformation. It has been said that the policy was to denigrate the virtues of the enemy and then you can represent him as the 'Devil Incarnate'.

The emotions directed against Priestley were almost certainly intended to encourage the people to lynch him. He narrowly escaped from Birmingham reaching Dudley with the help of friends. He then crossed the moors on horseback to Bridgewater and was lost but reached Kidderminster. He left by a hired chaise (a light carriage) for Worcester and took a mail coach to London where he was safe. The driver of the chaise told his friend that had he known that he was carrying the notorious Dr Priestley he would have turned the vehicle over in an attempt to kill him, no matter at what risk to himself. Thus, Priestley sacrificed himself for English democracy.

The King wrote to Henry Melville, the Home Secretary: "As the mischief did occur it was impossible not to feel pleased at its having fallen on Priestley rather than another, that he might feel *the wickedness of the doctrines of democracy* that he was propagating".

In 1884 the franchise in Britain was extended from 9% of adult males to 80%, and in 1913 it was extended to 90%.

May we remember Dr Joseph Priestley for his sacrifices for the ordinary, unprivileged people of Britain as the greatest Briton ever! Certainly in his home county.

Footnote

Antoine-Laurent Lavoisier died swiftly with his father and twenty seven other senior tax officials; Joseph Priestley saw his life's work destroyed and his reputation impugned by outrageous lies which have not yet been fully exposed and recognised to our profound shame.

PRIESTLEY AND SUBSCRIPTION LIBRARIES

Geoffrey Forster
Leeds Library

Geoffrey Forster

This is an outline of a paper showing the part played in the life of Joseph Priestley by books and libraries, an aspect of his life neglected previously. Books and libraries were not only important in Priestley's own education and development but were also crucial in disseminating his philosophy and work to as wide an audience as possible. In turn, he hoped that readers, pupils and members of his congregations would read widely for themselves and form their own informed opinions.

Joseph Priestley was born in 1733 at Fieldhead near Birstall in the West Riding of Yorkshire. When only about six years of age, his mother died and he went to live with his Aunt Sarah Keighley at nearby Heckmondwike. He proceeded to attend a number of local schools including Batley Grammar School. Books were already becoming an important part of his life. In his spare time, he read whatever was available (sometimes borrowing books from a family friend, a Mr. McGill) and by this means he was able to extend himself beyond the limited teaching available locally. He showed a particular talent for languages.

On finishing school, a career in commerce beckoned but an improvement in Priestley's often delicate health led his aunt to renew her previous hope of his becoming a minister of religion and he was sent to Daventry Academy. The dissenting academies like Daventry provided a university education for students excluded from the universities of Oxford and Cambridge because of their religious beliefs: the students included those training for the professions as well as those intending to be dissenting ministers. These academies generally had a philosophy that all sides of an argument should be available to their students who were then expected to make up their own minds. To this end, the tutors at Daventry recommended books to their students that, according to Priestley's *Memoir*, could all be obtained from the local *public* library (which was possibly the grammar school library). Additionally, each day, Priestley undertook the translation of ten folio pages in Greek with his friend, a Mr Alexander. A surviving portion of his journal shows the tremendous quantity of other reading he also undertook. One of the books that he came across – David Hartley's *Observations on man* – was to change his entire outlook with its argument that men are shaped by their circumstances and thus could be directed and perfected.

Books continued to be crucial to Joseph Priestley after he left Daventry in 1755. He was an avid collector of books for himself: these he was always happy to make available to his students or to members of his congregations. However, acquiring books was not always easy for him on limited stipends and cut off from the bookshops of London which he visited whenever he could. During Priestley's first ministry at Needham Market in Suffolk, he struggled to find books of science and his experimentation suffered as a result. However, the personal library of a local quaker helped to supplement his general reading. On moving to Nantwich in Cheshire in 1758, his reading was devoted to that needed for his work as a minister and later as tutor to a school: he had little time for reading either for leisure or general improvement.

Joseph Priestley left Nantwich for the Warrington Academy in September 1761 where he became tutor in languages. The academy at that time was in an old house on the side of the Mersey but in 1762 it moved into new purpose-built premises that provided an enclosed academic precinct for tutors and students.

The curriculum at Warrington was very narrow when Priestley arrived there. He introduced and taught several new subjects including history and general policy, the laws and constitution of England, the history of England, elocution, logic and Hebrew. One year, he even taught anatomy. To support his teaching, he produced and printed lectures for his students. Many of these were eventually published for more general consumption. He also began work on a general history of science: the volume on the history of electricity was completed in 1767 and was printed by William Eyres in Warrington. Perhaps his best known teaching aid and most successful publication was his massive *Chart of biography* which was again printed by Eyres. A nomination and election as a Fellow of the Royal Society was the direct result of the chart's publication.

It was in Warrington, that Joseph Priestley first became directly involved in the organisation and running of a library. He thereby played his first part in the revolution occurring in the availability of books and libraries in the British Isles between his birth in 1733 in Yorkshire and his death in 1804 in Pennsylvania, North America, In 1733, almost all the available libraries were attached to places of education or worship and had a specialised stock (usually in Latin). However, following the lapse of the Licensing Act in 1695, the increasing number of printing presses and the end of censorship had led to a massive increase in the number of newspapers, journals and books published. Additionally, these published items were for the most part in English making them accessible to a much wider audience. Books, though, were expensive and only the wealthy could afford to buy them in large numbers. Instead, it became common for booksellers to allow customers to borrow and return books for a small charge. But this did not answer everyone's needs. Such *circulating libraries* were invariably temporary – not intending to become permanent collections. And the first 'free' public libraries providing general collections paid for by local taxation were not to appear in Great Britain until the middle of the nineteenth century.

So the increasingly literate, prosperous and leisured members of the middle class (and in some places, literate members of the working class) from the mid-18th century onwards created their own *subscription libraries*. They combined their resources to create a library of mutual benefit where books on a wide

Statue of Joseph Priestley, City Square, Leeds, West Yorkshire.

Mill Hill Chapel today as seen from the eyes of the statue of Joseph Priestley looking across Leeds City Square.
PAINTING BY STEPHEN BARLOW 2004 FOR THE PRIESTLEY SOCIETY

range of subjects were acquired when new and were kept in perpetuity. New subscribers generally paid a one-off capital sum and an annual subscription. These libraries were run by the subscribers for their own benefit. They were especially popular amongst *dissenters* because their rejection of certain tenets of the established Anglican Church excluded them by law from the old established schools and universities and, as a result, from the most valuable libraries. It must not be overlooked, though, that members of the middle class generally would have found it difficult to gain access to such institutional collections.

In 1760, the Warrington Circulating Library had been established. Despite its name, this was a members' *subscription* library and not a bookseller's *circulating* one. It was based on a library founded in Liverpool in 1758. The library's founders were generally members of the academy. Its first president, John Seddon, was one of the academy's own founders and a friend of Priestley. Priestley himself joined the library's committee in 1762 on the resignation of a Mr. Eden and served on it until 1767 when he removed from Warrington due to his wife's poor health and the academy's poor progress. Unfortunately, insufficient records survive from Priestley's period in Warrington to paint a more detailed picture of his involvement at the library.

In September 1767, Joseph Priestley moved from Warrington to Leeds to take up his appointment as the new minister of Mill Hill Chapel. He was accompanied by his wife and daughter, The Priestley family's first home was out of town in Meadow Lane next to the Jaques and Nell Brewery where Joseph studied the action of the fermentation gases. In summer 1768, the family moved to a house in Basinghall Street (alongside Mill Hill Chapel). On 24th July 1768, a son – Joseph – was born.

On 9th August 1768, advertisements appeared in the Leeds newspapers for a proposed *circulating library*, again along the lines of the one established in Liverpool. Arrangements for the library were suggested and those interested in supporting the venture were invited to a meeting the following week. The library then established, now known as the Leeds Library, is, today, the oldest surviving of the proprietary type of subscription library – where members paid a capital sum to join (a guinea in 1768) and an annual subscription to retain membership (5 shillings in 1768). It opened on 1st November 1768.

It is highly likely that Joseph Priestley drafted the newspaper advertisement and compiled the first catalogue at the end of the year. Having had experience of running the Warrington Circulating Library and knowing members of the Liverpool Library, it is certain that Priestley was one of the prime movers (if not *the* prime mover) in the establishment of the library. He was to be its first secretary and second president. Many of the members of his congregation at Mill Hill Chapel were to be found amongst the 104 initial subscribers. These subscribers at first met in a first floor room above Joseph Ogle's bookseller's shop at the sign of the Dial at Kirkgate-end. Ogle himself was appointed the first librarian. The library was later to move to rented accommodation fronting Rotation Office Yard in Kirkgate (1781–1808) before coming in June 1808 to its present purpose-built home in Commercial Street.

Joseph Priestley left Leeds in 1773 to take up a position as librarian and companion to William Petty, 2nd Earl of Shelburne. Shelburne had been Secretary of State in Chatham's administration but was dismissed in 1768 for his opposition to government policy towards the American colonies. He effectively became Priestley's patron paying him a salary of £250 a year and promising an allowance for life should they ever separate. Priestley, though reluctant to leave Leeds, was always anxious to provide adequately for his family (the salary was twice that given to him at Mill Hill). In his *Memoirs*, Priestley states: "My office was nominally that of *librarian*, but I had little employment as such, besides arranging [Lord Shelburne's] books, taking a catalogue of them, and of his manuscripts, which were numerous, and making an index to his collection of private papers." Priestley also supervised the education of Shelburne's two young sons.

The Priestley family was given a house on Shelburne's estate at Bowood House at Calne in Wiltshire. Here Joseph was able to fit out a new laboratory and library and to continue his experiments on air. He went to London each year for the winter season and, during these visits, he toured the bookshops and libraries. In one bookshop, he made the acquaintance of John Adams, the future American President. In 1774–1775, Shelburne and Priestley toured Europe and, whilst in Paris, they dined with the French chemist, Antoine

Laurent Lavoisier. Lavoisier, dissatisfied with Priestley's *phlogiston* theory with regard to the composition of air, undertook work which was to lead finally to the discovery and naming of the element oxygen.

Priestley's years with Shelburne were productive ones in terms of his experimental work and printed output in the fields of both chemistry and theology. However, in 1780, the two men parted for reasons not entirely clear. It is possible that Priestley's political views were an increasing embarrassment to Shelburne: the Earl had also recently remarried and his new wife may not have taken to her husband's companion.

In September 1780, Joseph Priestley moved with his wife and family to Birmingham. His initial hope was to have more time to pursue his scientific and theological studies but, after getting over a bout of illness, his concern was once again how to provide for his family. He received many generous gifts of money from relatives, friends and other supporters and his brother-in-law, John Wilkinson, found him a house at nearby Fair Hill. It helped further when, in December 1780, he accepted an invitation to become the minister of Birmingham's New Meeting dissenting congregation.

Now recovered in health, more comfortably off and happily settled with a new congregation, Joseph Priestley's thoughts could occasionally turn to other pursuits and it was not long before he was involved with another library. When Priestley came to Birmingham, there was already a subscription library – the Birmingham Library – founded in 1779. This, though, had failed to make much of a figure and amounted to little more than a few members borrowing from a few shelves of books in the home of the father of one those members. Priestley joined and at once set about remodelling the institution on the lines of the more successful library at Leeds. In June 1781, he placed advertisements in the newspapers inviting subscribers to meet to consider the library's future. He redrew the library's rules on the lines of those in Leeds and he encouraged the creation of a separate *foreign* section as had occurred in Leeds in 1778. The library moved to Swan Yard where it was placed in the custody of a paid librarian. By December 1782, an ever increasing number of subscribers had access to over 500 volumes increasing just two years later to 900 volumes.

The subscription library in Birmingham as elsewhere provided more than just books. It not only brought people of a similar persuasion together – the Birmingham Library's members at first being almost exclusively dissenters as well as enthusiastic readers – it also provided a chance to develop connections with other members of local society in the way that other new institutions such as friendly societies, masonic lodges and assembly rooms did. Joseph Priestley, whilst in Leeds, had received enthusiastic support in his experimentation from another Leeds Library member, the Methodist surgeon, William Hey. In Birmingham, he became part of that renowned network of scientists, industrialists and artists, the Lunar Society (so called because the members of this society met monthly on the Monday evening nearest the full moon). Though based more widely in the Midlands rather than just in Birmingham, several of its members, along with Priestley, were subscribers to the Birmingham Library.

One might think that the reorganised Birmingham Library was now set fair. However, tensions within the town were soon to be both reflected in and exacerbated by library affairs. So much so that, in July 1791, when, during four days of rioting, a mob ransacked and destroyed the property of the Birmingham dissenters, a local commentator, William Hutton, suggested that disagreements at the town's subscription library were *a* major if not *the* major cause.

What had gone wrong? The Birmingham Library was started by Protestant dissenters but, during the 1780s, it came to be dominated by members of the local Anglican community. The division on religious grounds between the subscribers was reflected in conflict over the addition (or not) of works of political or theological controversy. Joseph Priestley had opposed such additions until the library could afford to purchase all sides of an argument. It was, though, to be the acquisition of his own controversial work, *An history of the corruptions of Christianity*, that provided the key flashpoint in library affairs and, thereafter, in the Birmingham newspapers and in locally published pamphlets. The division on religious grounds was also reflected in disputes regarding perceived attempts to load the library's committees with people of a particular religious persuasion.

The disputes at the library were not the immediate excuse for the Birmingham riots of 1791: it was rather a dinner held on the second anniversary of the fall of the Bastille – 14 July – at a time when concern about events in Revolutionary France was intense. An inflammatory handbill suggesting that those attending the dinner supported French republicanism may well have been a forgery but it served to provoke the mob still further. The so-called supporters of *Church and King* set out to teach the disloyal dissenters a lesson.

William Hutton, as well as arguing that the disputes at the library had helped to initiate local tensions, additionally cited the campaign for the repeal of the Corporation and Test Acts – the acts preventing protestant dissenters amongst others from participating in public life – as another key factor in the lead up to the riots. Joseph Priestley was at the forefront of the campaign for the repeal of these acts and was a target for verbal and printed attacks by local Anglican churchmen – it being said by Hutton that such attacks against him were the surest way to gain preferment in the Anglican church. To make matters worse, ill chosen words about putting gunpowder under an edifice of error and superstition had added to Dr. Priestley's local and national notoriety and had earned him the unwanted sobriquet of *Gunpowder Joe*.

During the riots, dissenting chapels and the houses of leading dissenters were systematically destroyed. Priestley's chapel – the New Meeting House – fell victim as did Priestley's house at Fair Hill. His laboratory, library and manuscripts were all lost. The family escaped and fled, never to return to Birmingham. Leading local Anglicans denied inciting a mob consisting, it was said, of people who "would have sold their King for a jug of ale, and demolished the Church for a bottle of gin". Involved or not, the rioters were left unchecked by local magistrates. They were eventually stopped by the intervention of the Nottingham militia. Coincidentally, the now Anglican-dominated Birmingham Library, on whose committee Priestley in 1791 declined to serve, was not attacked.

Following the riots, Joseph Priestley and his family moved from Birmingham to London where he obtained a ministry at the Old Gravel Pit Meeting House in Hackney. He additionally undertook some teaching at the dissenting Hackney College. The local people expressed concern at the arrival of this radical theologian in their midst fearing that the events of 1791 in Birmingham might be repeated there. The members of the Royal Society in London shunned him.

It is not known for certain whether Joseph Priestley used any libraries during his years at Hackney. There was a subscription library – the London Library Society – established in 1785 near St Paul's Cathedral. Leading dissenters were prominent amongst the early subscribers including some influential at the Hackney College, such as Abraham Rees and Andrew Kippis, and friends of Priestley, such as Adair Crawford, Benjamin Vaughan and William Vaughan. Few records survive for this library which possibly combined with another subscription library, the Westminster Library, and which had definitely closed by 1821. Nevertheless, it seems highly likely that Priestley would have joined and used it.

In 1794, Joseph and Mary Priestley left England for North America and Northumberland, Pennsylvania to rejoin their three sons who had departed there the previous year. In Pennsylvania, Joseph Priestley worked for the establishment of a Unitarian congregation but resisted the offer of an academic chair in chemistry at Philadelphia. He built a house (with library and laboratory, of course!) which is now the home for the Joseph Priestley House Museum. Sadly, his wife died in 1796 before the house's completion.

Priestley visited Philadelphia each winter throughout his time in America despite the journey being an arduous one. He was a member there of the American Philosophical Society (with its library) and it is known that he also visited the Library Company of Philadelphia. His own books had been shipped to Northumberland and were increased there by further acquisition. Given this lifelong love of books and libraries, it was appropriate that, when Priestley died on 6 February 1804 after a few years of declining health, he was lying peacefully in a cot in the library of his new home.

Today, the two subscription libraries, with whose foundation Joseph Priestley was most closely involved, still flourish – those in Leeds and in Birmingham. At each, Priestley's name is remembered fondly and his contribution is gladly acknowledged: the library at Birmingham, incorporated since 1955 with the Birmingham and Midland Institute, is now known as *The Priestley Library*; the Leeds Library has named a room after its most famous founder member; and both libraries hold lectures in his honour – the Leeds Library's in collaboration with Mill Hill Chapel, Priestley's old chapel.

Further reading

Beckwith, Frank – The beginnings of the Leeds Library, *Publications of the Thoresby Society*, 37 (Leeds, 1941), pp. 145–65.

– The eighteenth century proprietary library in England, *Journal of Documentation*, 3 (1947–1948), pp. 81–98.

– The Leeds Library 1768–1938, 2nd edition (Leeds, 1994)

Forster, Geoffrey, *et al.* – A very good public library: early years of the Leeds Library, *History of the Book Trade in the North*, 82 (Newcastle-upon-Tyne, 2001)

Gibbs, F. W. – Joseph Priestley : Adventurer in Science and Champion of the Truth (London, 1965)

O'Brien, P. – Warrington Academy, 1757–86: its Predecessors and Successors (Wigan, 1989)

Parish, Charles – History of the Birmingham Library (London, 1966)

Priestley, Joseph – Autobiography of Joseph Priestley: Memoirs Written by Himself, Compiled with Notes and an Introduction by Jack Lindsay (Bath, 1970)

JOSEPH PRIESTLEY – TO TRUTH, TO LIBERTY, TO RELIGION

Peter Hewis
Harris Manchester College, Oxford

Peter Hewis

I feel a great affinity with Joseph Priestley for several reasons.

Firstly, I share his birthday of March 13th although the calendar changed between his birth and mine!

Secondly, in 1991 I was privileged to make a brief speech at his statute in Birmingham; we were there to recollect the 200th anniversary of the Priestley Riots in that city.

Thirdly and most importantly, today I represent Harris Manchester College in Oxford, the successor to Warrington Academy. Joseph Priestley would surely be delighted that in recent years we have welcomed students with such diverse backgrounds as a navy chef, a railway signalman, a woman with the experience of a jail sentence and a retired High Court judge. Since becoming a full Oxford College, the college has expanded its number of courses and in so doing it has renewed links with Priestley's days at Warrington. For one thing we again have medical students as did Warrington; we have the Professor of Diabetes Research and another in Palliative Care. On the side of commerce we have a new link with the Said Business School.

Where does this link with Priestley?

We have to look at the meaning of the words Dissenting Academy. After the year 1662 all non-Anglicans were barred from the old Universities such as Oxford and Cambridge; it was difficult even to maintain local places of learning.

Priestley plaque at Harris Manchester College

Dissenting means those Protestants that dissented from the Church of England and it included Independents, most of whom eventually became Congregationalists and English Presbyterians, most of whom gradually became Unitarians. Alexander Gordon ("Addresses Biographical & Historical", pub. Lindsey Press 1922) wrote,

> ... They employed every effort in their power to debar Nonconformists from the exercise of the teaching profession, and to break up their schools ... By hurried moves from place to placed, Teachers sometimes managed to evade arrest, at least for a time.

It was only in 1734 that things changed a little, when proceedings against Philip Doddridge were stopped by order of George II. If Dissenters wanted their sons (and in those days it was only sons) to have any kind of higher education, how could that be achieved? The answer lay initially with individual ministers who set up Academies attached to their local Chapel. To name just a couple, Rathmel in Yorkshire founded by Richard Frankland was one of the earliest in 1669/70. Move on a few years and we hear John Jennings of Kibworth then Hinckley described by Doddridge as a man who encouraged absolute freedom of inquiry amongst his students. Doddridge continued in that tradition when he took over the academy at Kibworth. We sense the influence of those early academies on Priestley when he said years later,

> Truth can never have a far chance of being discovered, or propagated, without the most perfect freedom of inquiry and debate.

Priestley Window at Harris Manchester College

What about Academy?

As you will know the word has a fine connotation in Scotland but is less known in England. Calvin used the word to describe a seat of learning in Geneva which was not under a Papal Bull and called his university Academia. (See Gordon pp 71/2) Alexander Gordon tells us,

> To maintain an open door was vital to the very being of the old Nonconformist (or Dissenting) Academy.

The students covered many subjects, some trained for the legal professions, some for the medical profession, many for the Ministry. Doddridge brought in a great change and introduced lectures given in English, not Latin. Again Gordon describes three branches to the curriculum, Science, Theology and Philosophy with prominence being given to the latter. The aim of the academies was to equip the students for employment in many professions. You can sense the ethos in some mottos, that of Frankland at Rathmel was, "*'Tis no free country till the mind be free.*"

Even today in Oxford over our college door are the words, (in English not Latin) "*To Truth, To Liberty, To Religion.*"

Into this broad, thought provoking atmosphere Joseph Priestley entered when he was enrolled in 1751, age 18, at the newly opened Academy at Daventry. He enrolled in the spirit of free inquiry but his days in Daventry are not in my brief today.

Warrington Academy: 1761 to 1767

After two short ministries Priestley reached the age of 28 and became the tutor in Languages and Polite Learning at Warrington Academy, in the same year of 1761 he published *The Rudiments of English Grammar* and it remained in use for at least 50 years. As the Encyclopaedia Britannica notes, "a work that was radical in its reliance on description of actual English usage as opposed to prescription on the basis of dead classical languages." Within the next four years he produced books on Theory of Language, Chart of Biography, a Course of Liberal Education for Civil and Active Life, Lectures on History and General Policy and even a History of Electricity.

Instead of teaching the classical traditions of Oxford and Cambridge, Priestley emphasised history, science and the arts; his aim was to train students for the field of employment in a practical way. In his Memoirs he writes,

> It was my province to teach elocution, and also logic and Hebrew. The first of these I retained, but, after a year or two, I exchanged the two last articles with Dr. Aikin for civil law, and one year I gave a course of lectures in anatomy.

For greater detail of his time in Warrington you should read the Ph.D. Dissertation by Norman Rose, "Joseph Priestley: His Impact on and Achievements at Warrington Academy". I will just quote one passage from Norman's Dissertation because it illustrates how meticulous Priestley was. It comes from Priestley's "History of Electricity."

> The electrician, therefore, ought never to be without the common tools of a cabinet maker, clock, and watch maker, at least, and know, in some measure how to use them.

Artefacts from Priestley's days at Warrington Academy, now in the custody of the librarian at Harris Manchester College, Oxford. The items include Priestley's clothes brush, two pairs of spectacles, a legal document with family seal and other medals and correspondence

With respect to glass, he ought by all means, to learn the use of a blow pipe, the method of drawing out and bending glass tubes, and performing, with some degree of dexterity, other operations on glass, which he will want to use in a great variety of forms. An electrician, thus furnished, will be able, upon any occasion, to serve himself: and the slowness and blunders of mechanicks do but ill suit with the ardour of persons engaged in philosophical inquiries ... Formerly, indeed philosophers were obliged to construct their own machines. Mr Boyle, Mr Hawkesbee and Doctor Desaguliers would have done nothing by giving tradesmen order for what they wanted.

If you want more such gems then you will be most welcome to see our many Priestley archives at Harris Manchester College.

In just five years Priestley's time made the name of Warrington famous throughout England and probably beyond as Edinburgh University awarded him the degree of LL. D in 1765. According to V. D. Davis in his *A History of Manchester College* (1932) the Academy was founded, "on more public, broader lines than the Academies which preceded it".

If you know the Bridge Foot today by the River Mersey in Warrington then you will find it surprising that Anna Laetitia Barbauld could describe the students and their surroundings in these terms, quoted in V.D. Davis, pp 33/34.

Ye generous youth who love this studious shade,
How rich a field is to your hopes display'd!
Knowledge to you unlocks the classic page,
And virtue blossoms for a future age.

The days at Warrington saw Priestley making a massive contribution to the education of his day.

Priestley affects education today

We can sense the Priestleyian influence on education in a letter which Priestley wrote to the then Prime Minister. On March 31st 1787 whilst just a few miles from here, in Leeds, Priestley wrote to, William Pitt regarding education at Academies like Warrington then Manchester and contrasting it with the education given at Oxford and Cambridge. Here is a little of the letter.

> But though ignominiously, and unjustly, excluded from the seats of learning, which, as maintained by the public funds, ought to be open to all the community and driven to the expedient of providing, at great expense, for scientific education among ourselves, we have the advantage, that our institutions, being formed in a more enlightened age, are more liberal, and therefore better calculated to answer the purpose of a truly liberal education. Thus while your universities resemble pools of stagnant water, secured by dams and mounds and offensive to the neighbourhood, ours are like rivers, which taking the

> natural course, fertilize a whole country. Our plans of education embrace a much greater variety of objects; and the minds of our youth, being unfettered by subscription, are certainly more open to the impression of truth.

Imagine a Prime Minister receiving that letter; imagine the stuffy Fellows of Oxford reading those words. Priestley was challenging the Government and the educational establishment of the day. He was pleading for them to widen their outlook, to become open to students from all backgrounds. He was a pioneer, leading on (in my view) to the establishment of the red brick Universities, eventually leading to the abolition of the religious tests of Oxford and Cambridge. I also have a feeling that Priestley's thoughts led well over a century later to our state education system.

In the brief years at Warrington only around four hundred students were enrolled; they studied for the Law, Medicine, the Army, Trade and fifty five of them were Divinity students; surprisingly thirteen entered the C of E and one became an Irish Bishop. Priestley's time at Warrington achieved far more than teaching students, writing books and undertaking experiments; it possibly changed the whole course of English, maybe even the whole of British education. Writing in 1906 Robert Spears "Memorable Unitarians" said of him, "He was a friend of popular education and the defender of liberty of conscience all the world over."

Let me end with part of the wording on a memorial to Priestley at Unitarian New Meeting, Birmingham.

> His example as a Christian will be instructive to his wife and interesting to the Good of every Country and in every Age. This tablet is consecrated to the memory of the Reverend Joseph Priestley by his affectionate congregation.

Joseph supported the education of women and said,

> Certainly the minds of women are capable of the same improvement and the same furniture, as those of men.

Well, he was instructive to his wife and he has certainly been instructive to the world of education. For a Lancastrian to admit that of a Yorkshireman is no mean feat and so on March 13th I shall drink a toast to Joseph Priestley.

JOSEPH PRIESTLEY: EDUCATION FOR A NEW SOCIETY

Ruth Watts
University of Birmingham

Ruth Watts

Joseph Priestley, a great creative scientist and radical theologian and politician, was also a progressive educationalist with a profound influence on educational developments in England. His environmentalist and rational philosophy of education was underpinned by his experience firstly as a schoolteacher, then as an innovative lecturer in an astonishing range of subjects at the liberal dissenting academy of Warrington, and from 1791 to 1794 a lecturer in science and history at Hackney Academy in London.[1] Here his educational philosophy will be briefly examined and then its significance investigated for its effect in Birmingham where he lived from 1780 to 1791.

Priestley's educational philosophy

People, Priestley insisted, were not what they were born but what their education had made them. A careful education from birth in all aspects of being – intellectual, moral and physical – was a prerequisite, therefore, for the virtuous, useful and happy man or woman. Priestley derived his ideas partly from John Locke who was admired by many dissenting educationalists and, indeed, others, but even more so from David Hartley whose *Observations on Man* he reissued in condensed form. Following Hartley's associationist psychology which maintained that all complex or 'intellectual' ideas arise from simple ones, formed in turn from external impressions made on the senses,[2] Priestley eagerly proclaimed that people developed through individual associations and circumstances[3] – "children may be formed or moulded as we please".[4]

This educational philosophy/psychology became the basis of Unitarian education and integrally intertwined with Unitarian religion. Reflection, experience and extensive intellectual education were seen as the prerequisites for individual moral, religious and intellectual progress and even perfection, not innate cause or divine intervention.[5]

Thus it appeared that all people of both sexes should receive the same, careful, wide education and that parents and teachers in particular should fully understand associationist psychology and be well-educated themselves. This education should be scientific in method and in content. Thus empirical and experimental subjects such as modern history and science were extolled, both new subjects in higher education and in schools.[6] Priestley's own teaching and writings also helped the acceptance of English and its literature as an academic subject.[7] He anticipated the use of sources in history teaching and of experiment in science teaching way ahead of his time, adding that all studies should be adapted to the age and the capacity of the learner.[8]

Priestley, indeed, saw particular kinds of knowledge as power – power to create virtuous beings and to control the forces of nature for good. To him science was that paradigm of free enquiry which was of "peculiar concern" to his religious group, the Unitarians and fitted his perception of the needs of the rising industrial and commercial middle-class in which many dissenters, including the energetic Unitarians, were to be found.[9] Certain that the leaders of the future would come from those who had mastered the sources of knowledge which had changed the world, he perceived that the very scientific and industrial interests which were scorned in the traditional education of Oxbridge and the public schools were the just basis of a prosperous meritocracy and a proper part of a "truly liberal education" which would produce enlightened leaders of the middle-class.[10] Thus through his teaching and prolific writings he urged a new liberal and useful education and became a dramatic and innovative force in education. He greatly influenced both the Unitarians and the Utilitarians, each leaders in English educational reform from the late eighteenth through the nineteenth centuries.

Priestley and educational reform in Birmingham

In Birmingham Priestley had particular effect on the Unitarians through his preaching at the New Meeting and teaching in the Sunday school; through his part in the setting up of the library; and, on all his family and friends particularly those in the Lunar Society.[11] From 1780 to 1791 he was a leading light in the latter, a small but vital Society which included Matthew Boulton, James Watt, Josiah Wedgwood (another Unitarian) and others. This dynamic group of creative scientists, enterprising inventors and business men and innovative writers also had educational significance.[12] The Lunar Society itself was an educative society, in the way that its members eagerly exchanged ideas, excitedly caught up and developed each others' suggestions and enthusiastically engaged in a range of subjects upon which they poured their increasing expertise. Believing in a new modern and scientific education, they showed their advanced ideas through the education they gave to their own children.[13] Furthermore Joseph Priestley, Richard Edgeworth, Thomas Day, Erasmus Darwin (plus Maria Edgeworth since she wrote with her father), all published very influential educational works.[14] Like Priestley, their revolutionary educational idealism was based on their conviction that the new sources of knowledge which were changing the world, together with literary excellence and moral development, formed a "truly liberal education". Such, in turn, would forge a new type of gentleman – enterprising, open to new ideas, tolerant, humane, liberal and civic-minded unlike what they saw as privileged, duelling, pleasure loving selfish aristocrats. The 'Lunaticks' were not against the arts but fused them with science to produce a new culture, as was exemplified in Wedgwood's pottery and Darwin's poetry.[15] This rational, useful, scientific, liberal, (middle-class) education was revolutionary for its time, however, and hugely contested during the long wars with France.

Nevertheless, Priestley's teachings had a lasting effect. In Birmingham this occurred particularly through the Unitarians (rational dissenters who rejected both the trinity and original sin), who met in Priestley's chapel, the New Meeting and its sister the Old Meeting. The direct influence of Priestley was extolled by William Matthews who had become a Unitarian after hearing Priestley "pouring out ... the great stores of his most capacious mind to a considerable number of young persons of both sexes, whom ... he encouraged to ask him questions ..."[16] Various societies, including one giving a more extensive education to youths and lectures on science and mechanics to factory workers, Thomas Clark's scientific lectures to the 'cast-iron philosophers' and the Brotherly Society which trained teachers from the boy pupils of the Sunday schools, were all recalled with great pride by their participants. Other ventures such as the 'Philosophical Institution' founded in 1800–1, where men like Thomas Wright Hill lectured on science, led Matthews in 1827 to laud Birmingham "as one of the most striking instances and strongest proofs of the civilizing and moral effects of education, that characterize modern times".[17]

Such ventures, then and now, have been criticised for their paternalistic attitudes and certainly there was a gulf between the seemingly egalitarian activities of the Brotherly Society and the increasingly wealthy and professional members of the Old and New Meetings. Yet the Brotherly Society did help educate men like James Luckcock, later prominent in progressive Unitarian Sunday school education and a prosperous manufacturer, and Thomas Wright Hill who with his wife Sarah Lea opened Hill-Top, a secondary school for boys, in 1803. Hazelwood, the successor to this in 1819, has been termed an 'educational refraction of Priestley's ideas'.[18]

The Hill children received their education principally through constant eager discussion and debate at home, self-education and, for the five brothers, the hard discipline of teaching from an early age in the family schools. It was Rowland, however, who was the prime architect of Hazelwood, the principles of which were published by him and Matthew in their 1822 treatise which became known as *Public Education*. Hazelwood became a brilliant showplace of almost revolutionary ideas in education but the Hill family always acknowledged their debts to others and of these the principal were Priestley and Maria Edgeworth.[19]

Hazelwood's aim of useful, pupil-centred education which would give its scholars sufficient knowledge, skills and understanding to allow them to continue self-education through a life "most useful to society and most happy to himself",[20] fitted Priestley's ideals admirably. The Hills extolled the skills of good, teaching which stimulated "clear, vivid and accurate conceptions" through the association of ideas through drama, costume, visual aids and gave learners the encouragement of repeated success. Their whole educational practice illustrated associationist ideas and their curriculum reflected their enthusiastic delight in practical mathematics and science although they were too astute to allow science and technology to dominate.[21]

Furthermore, the whole system was designed to produce order, self-discipline, initiative and self-activity, those qualities indeed vital to an energetic, entrepreneurial, successful commercial middle-class.[22] The teaching and, indeed, the entire system, encouraged a large amount of pupil participation. Such principles made Hazelwood a regular showplace for distinguished visitors who longed for the reform of middle-class schooling,[23] yet the middle-class generally was largely unappreciative of the schoolboy republic and suspected the owners' political liberalism and undogmatic religion. Although Bruce Castle, a less radical offshoot of Hazelwood in London, was headed from 1833 by Arthur Hill for thirty-four years, Hazelwood closed in 1833 and therefore could hardly become the basis of a national system as its founder had so deeply desired.[24]

Nevertheless, the school was responsible for educating many of Birmingham's nineteenth century elite, 'gentlemen' in Birmingham at least. The Hills themselves, united in all they did, were extraordinarily prolific in invention and reform of all kinds including Rowland, inventor of the penny postage stamp. Local pupils included an impressive array of inventors, scientists, liberal politicians and local leaders including a number of the twenty-three Unitarian mayors of Birmingham between 1841 and 1893. The Hills and their ex-pupils were engaged in a whole plethora of social and political reform, unsurprisingly really since their older relatives had been supporters of the French Revolution along with Priestley in the 1790s and some had suffered with him. They all knew each other and were intermarried. Many were Unitarians.[25]

Such inter-connections helped these families to fight for educational and other reforms together although it proved difficult to establish either dissenting rights in public education or a liberal, modern scheme of education for the commercial middle-classes.[26] Generally such ventures were male orientated but evidentially both Unitarian homes and those of the Lunar educationalists were characterised by an easy intellectual equality between men and women.[27] Some women used their better education professionally as the example of Priestley's granddaughter Catherine Finch illustrated. She ran a school for girls in Edgbaston which Unitarian girls attended. She was reputed to be an 'advanced teacher', writing out her own books for her pupils and using excellent methods of teaching particularly in geography. Her own great interests were in astronomy which she taught through cards, pounding holes for the stars, and conchology from which she left a collection to the museum for elementary schools.[28] Other Unitarians sent their daughters to Miss Byerley's school in Warwickshire, an outstanding school run by the granddaughters of Josiah Wedgwood.[29] The limitations on what they could do with such an advanced education in the nineteenth century were very real but at least in these circles women and men shared equally ideas, commitment and enthusiasm. Some went further. Frederic Hill and his wife, Martha Cowper, like Matthew Davenport Hill supported Barbara Leigh Smith and Bessie Parkes, great-granddaughter of Joseph Priestley, in 1854 in a move for women's rights.[30] This was the beginning of the Women's Movement which gradually did so much to alter perceptions of and opportunities for women.

It would be foolish to attempt to trace Priestley's influence too far into the nineteenth century. Nevertheless, in education, in science and technology, in widening the opportunities of both men and women below the upper classes, it can be seen that he, his friends and followers did have deep and significant effect on education for a new society.

NOTES

NB. All books published in London unless otherwise stated.

1. See especially J. Priestley, 'Introductory Essays to Hartley's Theory of the Human Mind' (1790; 1st ed. 1775) in *The Theological and Miscellaneous Works of Joseph Priestley* ed. J. T. Rutt, 25 vols. (1817–31) [hereafter termed *Works*] vol. III, pp. 167–96. See also R. Watts, 'Joseph Priestley (1733–1804)', *Prospects, Thinkers on Education*, (1995, UNESCO) vol. XXIV, no. 3, pp. 343–53; 'Joseph Priestley and Education', *Enlightenment and Dissent*, (1983) no. 2, pp. 83–100.
2. D. Hartley, *Observations on Man*, (New York, Delmar, 1976) I, p. 65.
3. Priestley, 'Hartley', p. 184.
4. J. Priestley, 'The Doctrine of Philosophic Necessity Illustrated' (1782) *Works*, III, 521; Hartley, *Observations*, 82, II, p. 453.
5. J. Priestley, 'Preface and Dedication to Heads of Lectures on a Course of Experimental Philosophy' (1794) *Works*, XXV p. 389; J. Priestley, 'Philosophic Necessity', p. 515.
6. J. Priestley, Lectures on History and General Policy (Philadelphia, 1803), *Works*, XXIV; 'The History and Present State of Electricity, with original experiments' (1767) *Works*, XXV; 'Experimental Philosophy', p. 385.
7. J. Priestley, 'A course of lectures on oratory and criticism' (1777); 'The Rudiments of English Grammar' (1798) Works, XXIII, pp. 257–482, 3–118.
8. Priestley, 'Oratory', p. 259; Miscellaneous Observations relating to Education (1780) Works, XXV, p. 219; 'History', pp. 54–202, 463–83; *A Familiar Introduction to the Study of Electricity*, Johnson and Payne, 1769, p. 10.
9. Priestley, 'Oratory', p. 255.
10. Priestley, 'Miscellaneous Observations, 5–228; 'History', pp. 5, 11, 22, 313–17, 403–15, 471–5; Electricity', 345; B. Simon, *The Two Nations and the Educational Structure* 1780–1870 (1974), pp. 84–7.
11. Priestley, A particular Attention to the Instruction of the Young ... Gravel-Pit Meeting in Hackney December 4th, 1791, *Works*, iv–ix and Preface; J. Uglow, *The Lunar Men* (2001), pp. 319–20, 406–7.
12. Simon, *Two Nations*, pp. 17–71.
13. Julia Wedgwood *The Personal Life of Josiah Wedgwood the Potter* 2 vols. (Manchester, 1915), II, pp. 433, 547–548, 555–6; Henrietta Litchfield (ed.), *Emma Darwin. A Century of Family Letters* 1792–1896 2 vols. (1915), I, pp. 61–2.
14. Maria & R. L. Edgeworth, *Practical Education* 3 vols. (1801); Thomas Day, The

History of Sandford and Merton, (18 ...; 1st ed. 1783) Erasmus Darwin, *A Plan for the Conduct of Female Education in Boarding Schools* (Derby, 1797).

15. E. Darwin, *The Botanic Garden*, (1795) canto 1, 84–90, note XXII, pp. 53–9 and passim.

16. William Matthews, *A Sketch of the Principal Means which have been employed to ameliorate the Intellectual and Moral Condition of the Working-Classes in Birmingham* (1830), pp. 6–14.

17. *Ibid.*, pp. 14–18, 22–5.

18. W. H. G. Armytage, 'The Lunar Society and its Contribution to Education', *University of Birmingham Historical Journal*, (1967–8) V, p. 67.

19. Rowland & George Birkbeck Hill, *The Life of Sir Rowland Hill* (1880), pp. 19–34, 49–50, 52ff., 142, 160–85; Rosamund & Florence Davenport Hill, *The Recorder of Birmingham.* A Memoir of Matthew Davenport Hill (1878), pp. 7–13; Constance Hill (ed.), *Frederic Hill. An Autobiography of Fifty Years in Times of Reform* (1894), 26–7; M. D. Hill & Rowland Hill, *Plans for the Government and Liberal Instruction of Boys in Large Numbers Drawn from Experience [Public Education]* (1822), pp. 121–2, viii, 105, 128, 192, 199, 204 ftn.

20. Elie Halevy, *The Growth of Philosophic Radicalism* (1972) Faber, pp. 153–4, 249–478, 433, 491.

21. M. D. & R. Hill, *Public Education*, passim; R. & G. B. Hill, *Rowland Hill*, pp. 91- 9; *Hazelwood Magazine* I, No. I, pp. 3–5; No. 13, pp. 1, 5.

22. M. D. & R. Hill, *Public Education*, passim.

23. *Ibid.*, passim; Jeremy Bentham, *Chrestomathia* (1816), passim; *Westminster Review* (1824), 1, pp. 75–9.

24. P. W. J. Bartrip, 'A Thoroughly Good School', *British Journal of Educational Studies*, (1980) XXVIII, 49–59; William Henry Ryland (ed.), *Reminiscences of Thomas Henry Ryland* (Birmingham, 1904), pp. 24–8, 77–8.

25. *Ibid.*, I, pp. 25–6, 69–70, 102–6, 184–98, 207–14 & passim; Colin C. Hey, *Rowland Hill: Genius and Benefactor 1795–1879*, (1989) 25–43, 175–6, passim.

26. E.g. *ibid.*, pp. 79–80; Herbart New, *Centenary of the Church of the Messiah (formerly New Street) Sunday Schools* (Birmingham, 1888), pp. 14–20; Matthews, Sketch, pp. 29–33.

27. E.g. Ryland, *Reminiscences*, passim; Maria & R. L. Edgeworth, *Practical Education* passim

28. Ronald Martineau Dixon, 'Priestley's Daughter and her Descendants', TUHS (1931–4), V. pp. 43–66, 288–93, 411–16; Norah Byng Kenrick (ed.), *Chronicles of a Nonconformist Family. The Kenricks of Wynne Hall, Exeter and Birmingham* (Birmingham, 1932), p. 145.

29. *Ibid.*, pp. 161–5, 217; Phyllis D. Hicks, *A Quest of Ladies. The Story of a Warwickshire School* (Birmingham, 1949), passim; DNB, 'John Kenrick', XXXI, pp. 14–16.

30. R. and F. D. Hill, *Recorder of Birmingham*, pp. 114–5; C. Hill, *Frederic Hill*, pp. 77, 92–9, 188–93, 233, 305–9; R. Hill, *Rowland Hill*, pp. 19–30, 47, 82, 142, 184–98; Dixon, 'Priestley's Daughter', pp. 53–4.

JOSEPH PRIESTLEY – UNITARIAN MINISTER

Ann Peart
Unitarian College, Manchester

Ann Peart

Joseph Priestley was minister to five congregations during his lifetime, but can only said to have been a Unitarian minister to just three of them. His religious views changed dramatically over the years; his friend Anna Laetitia Barbauld said of him that he "followed truth as a man who hawks follows his sport – at full speed, straight forward, looking only upward, and regardless into what difficulties the chase may lead him" (quoted in Ellis, 1874, Vol I, p. 36). Priestley was brought up by his aunt to be a Calvinist, but soon questioned the doctrine of original sin, and preferred the Arminian teachings.

It should be remembered that, although the Toleration Act of 1689 had led to the founding of dissenting congregations, it was still illegal to deny the Trinity (Ditchfield, 1968, pp. 46–47). This had forced those who espoused Unitarian ideas to avoid proclaiming them publicly (Gordon, 1895, p. 25). As late as the 1790s, when the sedition laws were revived in response to the violence of the French Revolution, Unitarian leaders such as Thomas Fyshe Palmer, Gilbert Wakefield and the publisher, Joseph Johnson were imprisoned or deported. It was not until the Trinity Act of 1813 that Unitarian beliefs were made lawful (Wilbur, 1969 II; Bolam et al., 1968).

Priestley's early interest in theological matters led him to train for the dissenting ministry at the Daventry academy, where open questioning was encouraged. His first ministry, at Needham Market in Suffolk (1755–1758), was not a success. His increasingly heretical views on the nature of Christ coupled with a speech impediment – he stammered – left him distressed and depressed. He moved to Nantwich in Cheshire, where he spent three happier years. Perhaps the success of the school for both boys and girls, which he ran alongside his ministry helped, but his marriage to Mary Wilkinson must have been a major contributor to his well-being, and possibly also the gradual overcoming of his speech problem. It was partly this difficulty that led him to spend six years out of the ministry, as a tutor at Warrington Academy where Arian views were widely held among the staff. Up to this point, Priestley still considered himself to be Arian in belief, and it was not until his next move that he described himself as Unitarian (Clark, 1990. Williams, 1990).

In 1767 he left the academy to be minister of the congregation at Mill Hill Chapel, Leeds. Here his Unitarian views were acceptable, if not strictly lawful, and the stipend sufficient to support a growing family. He was, he said, "very happy with a liberal, friendly, and harmonious congregation, to whom my services (of which I was not sparing) were very acceptable" (Schwartz and McEvoy, 1989, p. xii). He proved himself a practical and successful minister, and with the aid of considerable self-discipline, a gruelling daily timetable, and the ability to write amid disruptions from visitors, he continued his educational and scientific work alongside his ministry. However, after six years he again took a break from ministry, only to return in 1780 as minister to New Meeting, Birmingham, where he thought the congregation "the most liberal of any in England" (Schwartz and McEvoy, 1989, p. xiv). Here he relished in the intellectual life both within the Unitarian community and in the wider region for over ten years. He was a member of the Lunar society, where at the full moon he met with likeminded men to discuss inventions and discoveries of the day, and until the backlash resulting from the violence of the French Revolution brought rioting and destruction to his own meeting house and home (Schwarts and McEvoy, 1989, p. xiv: Watts, 1996, p. 353). He and his family escaped just in time. They took refuge with friends in London, and at Hackney Priestley conducted his last ministry in England. For three years he preached to "unusually crowded audiences" and taught the young people. Then he sailed for America, where he hoped to escape the fear of attack or persecution (Schwartz and McEvoy, 1990).

Priestley's philosophy and theology have been described often. His attention to the practicalities of ministry is less well known. He was concerned that some rational dissenting congregations had become mere preaching stations, where a minister was paid by his hearers to harangue them once a week, "the performance of the minister being the only object of attention, if they dislike his sentiments or delivery", they go elsewhere (Clark, 1990, p. 60). Instead he was concerned to build up the religious community. He developed a devolved style of leadership, with elders and deacons taking on responsibilities. The elders should meet as a consistory, what we should now term the committee, about once a month unless called on other occasions by the minister, "to consult together concerning the state of the church, and the best method of promoting its real interests; but let them has no power to proceed further than an admonition, without the consent of the people at large" (Clark, 1990, p. 61). There should be a record kept of public proceedings, and a list of the names of church members. Deacons were younger men who served communion, and took up collections to relieve the "industrious and helpless poor", and the elderly (Clark, 1990, p. 61).

Group dynamics were important, and members should be involved. He asked, "Was it ever known that any man became attached to any kind of society in which he was a mere cipher, in which he had no sphere of action, and when nothing he did had any chance of being heard of, or of being mentioned to his praise, either in the society or out of it?" Even listening to sermons was an active, communal act. "If any impression is to be made upon the heart and affections, if men are to be led to feel, in order to determine and act, the discourses will be delivered to much greater advantage to a society than to separate individuals ... every individual of the audience will feel the force of the exhortation much more sensibly himself, when he perceives the effect upon others" (Clark, 1990, p. 61). People were not to be passive listeners, but rather active responders and charitably critical. He exhorted members of the congregation "to attend, and to judge with candour, to embrace the truth whenever you perceive its evidence, and to exercise forbearance towards myself, as well as towards all other Christians, whenever you see reason to withhold your assent" (Clark, 1990, p. 61). Yet more was demanded, mere assent was not enough: "Our greatest duties relate not to speculation, but to practice. We are all hearers of the word; but the greatest article of all is to be doers of work assigned us by God to do" (Clark, 1990, p. 61).

He rejected the idea of being a "pastor" to a flock of sheep, but wanted a congregation free to "call in question received opinions".

Priestley's ideas on the qualities and education needed for ministers were ambitious. Ministers should have a good education in biblical languages, liberal science, modern literature, moral and metaphysical subjects, sermon composition and public speaking. It was this training, not any other authority, or ordination, that made a man a minister. Laymen could lead public worship. The sermon was not essential. Reading the scriptures, praying and singing were important because they are more emotionally motivating, and need no learned leader. Lay people could preside at services including baptisms and funerals (Clark, 1990, pp. 61–62).

However, Priestley did not consider himself a good leader: "The man who takes the lead in any body of men whatever must be a man who gives no offence, by openly espousing obnoxious sentiments; but whatever he may think privately, must be so much on his guard, as not to be capable of being directly convicted of heresy; or if he cannot conceal his leaning to any unpopular turn of thinking, he must always speak or write with the greatest moderation, so as to give the least offence that the case will possibly admit of. Now it is very proper that there should be such men in the world. I esteem and value them, thinking them very useful in several situations, but it is evident that I have not been one of them" (Clark, 1990, p. 138). He considered that he had written with "offensive freedom; with what regard to truth, and with what success, I leave to others and to time to determine" (Clark, 1990, p. 138).

References

Bolam, C. Gordon, Goring, Jeremy, Short, H. L., and Thomas, Roger (1968) *The English Presbyterians: From Elizabethan Puritanism to Modern Unitarianism*, London: George Allen and Unwin.

Clark, John Ruskin, (1990) *Joseph Priestley; A Comet in the System*, San Diego, CA: John Ruskin Clark.

Ditchfield, Grayson M. *Some Aspects of Unitarianism and Radicalism* 1760–1810 (PhD thesis, Cambridge, 1968).

Ellis, Grace A. (1874) *A Memoir of Mrs Anna Laetitia Barbauld, with Many of her Letters. A Selection from the Poems and Prose Writings of Mrs Anna Laetitia Barbauld*, Boston: James. R. Osgood & Co.

Gordon, Alexander. (1970, first published 1895) *Heads of English Unitarian History*, Bath, Cedric Shivers Ltd.

Priestley, Joseph (1817–32) *Theological and Miscellaneous Works*, (Ed. J. T. Rutt) 26 Vols, Hackney, London: G. Smallfield.

Schwartz, A. T. and J. G. McEvoy (Eds). (1990) *Motion Towards Perfection: The Achievement of Joseph Priestley*, Boston: Skinner House.

Watts, Michael R. (1978, 1996). *The Dissenters*, Oxford: Oxford University Press.

Wilbur, Earl Morse. (1952, reprint 1969) *A History of Unitarianism in Transylvania, England, and America*, Boston: Beacon Press.

Williams, George H. (1990) *Joseph Priestley: The Minister, Citizen, and Church Historian*. In Schwartz, A. T. and J. G. McEvoy (Eds). (1990) *Motion Towards Perfection: The Achievement of Joseph Priestley*, Boston: Skinner House pp. 93–107.

JOSEPH PRIESTLEY HOUSE: THE TRANSFORMATION OF A PRIVATE HOME INTO A PUBLIC HISTORY MUSEUM

Andrea Bashore
Joseph Priestley House, Northumberland, Pennsylvania

Andrea Bashore

The Joseph Priestley House is located at 472 Priestley Avenue, Northumberland, Pennsylvania, near the junction of the North and West branches of the Susquehanna River, approximately 65 miles north of Harrisburg, Pennsylvania. The historic site, a property of the Commonwealth of Pennsylvania administered by the Pennsylvania Historical and Museum Commission since 1961, features the 1798 Georgian-style home of the English-born theologian, educator, political theorist and natural philosopher, Rev. Dr. Joseph Priestley and his family.

The record of Joseph Priestley's life and work in England from 1733 to 1794 has been well-documented and published by numerous historians and Priestley enthusiasts in the past. Much has also been written about Priestley's years in America from 1794 to 1804. My focus for this presentation will be the process whereby Priestley's 1798 private home in Northumberland, Pennsylvania was transformed into an historic site and museum, National Historic Landmark, and National Historic Chemical Landmark dedicated to preserving and interpreting Joseph Priestley's legacy for present and future generations.

The Priestley House has a five-bay façade with central hall and a modified Georgian floor plan. In her design for the house, Mary Wilkinson Priestley, Joseph's wife, located a pantry for storage, stairs from the kitchen to the

Joseph Priestley House, Northumberland, Pennsylvania

basement, and an interior staircase to the second floor in place of a fourth formal room on the first floor of the house. The symmetrical wings contain the main kitchen on the left and Joseph's scientific laboratory on the right. Each wing has a sky-lighted storage space above it. The second floor stair hall is enhanced by light from the Palladian window over the public entrance door. There are two like-size bedchambers, a third bedchamber with access to a dressing room, and a smaller fourth chamber – possibly a storage room or another bedchamber. With few exceptions, each room on the first and second floors was heated by its own fireplace hearth in Priestley's day.

The third floor attic is an unheated space composed of an open storage area and three chambers that may have served as sleeping quarters for the family's servants. From within a corner of the storage area, narrow stairs turn and rise through the loft above and provide access to the roof-top observation walkway.

The Priestley House is often described as a Federal-style home by Americans who visit the site. Federal architecture takes its name from the Federal Period (ca. 1780 to 1830) in American history, an era when the country's founding fathers worked to formally establish the United States. They created a national style and identity reflective of the classical ideals of ancient Rome, a city whose ideologies were used as a model for American government, education and the arts.

Federal architecture is defined as a style that retains the symmetry, five-bay façade, and central hall floor plan of the earlier (ca. 1725 to ca. 1775) Georgian style, yet features a refined simplicity and lightness of ornamental details. To Joseph Priestley, his American home may have seemed a compromise – a British architectural style simplified and altered in appearance by the use of wood, a readily available building material, for its construction and exterior clapboard siding.

We will now look at how the Priestley House came to be built in Northumberland at the close of the eighteenth century and how it survived on the Pennsylvania landscape until today.

Priestley's Immigration and Early Days in Northumberland

When Joseph and Mary Priestley sailed from England for America in April 1794, it was almost three years after the Birmingham Riots during which a loyalist mob opposed to Priestley's libertarian political ideals and his nonconformist religious beliefs destroyed the family's personal possessions, household goods and Fair Hill, the house where they resided. In the opinion of some Englishmen, the outspoken minister and natural philosopher had become a symbol of danger because of his support for the revolution in France and a threat to the British Church and King. Persecution of dissenters in this fashion was not uncommon in England during Priestley's lifetime.

Priestley and his wife left their only daughter, Sarah, Mrs. William Finch, in England with her husband and children, and set out to join their three sons in the village of Northumberland, approximately 150 miles northwest of

Philadelphia. There, Joseph Jr., William, and Henry, along with their friend Thomas Cooper established headquarters for a land speculation scheme to develop a settlement designed to offer safe haven and repose to other English dissenters. The hoped-for community was proposed for a rugged location 50 miles to the northeast of Northumberland near Pennsylvania's Loyalsock Creek, but the territory proved difficult for the Priestley brothers and Cooper to develop and efforts for the community's advancement were soon given up.

Early in June, the ship *Sansom* docked at the port of New York with Joseph and Mary aboard. They were welcomed in New York, travelled on to Princeton, visited Philadelphia, and finally arrived in Northumberland in July. Priestley's early days in the village were occupied with financial concerns, Mary's health, his daughter Sarah's welfare, land acquisition, and thoughts of building a house – topics he addressed in numerous letters.

Priestley's pattern of activities in Northumberland was similar to his daily round in England. He preached sermons, walked in the village, engaged his neighbours with conversation, and undertook business transactions. He also penned correspondence to obtain laboratory apparatus, supplies and books, and worked to bring formal educational opportunities to inhabitants of Pennsylvania's rural countryside. At times, he lamented his inability to conduct experiments due to lack of suitable space and delays with receiving his equipment which was often damaged in shipment.

Priestley returned to Philadelphia, the nation's capital, on visits several times. There, he preached sermons at the Universalist church and organized a congregation of Unitarians. John Adams and many members of Congress attended a service at the church on February 14th, 1796 and heard Priestley's sermon. That same year, he was invited to take tea with George Washington in the city. Priestley also attended meetings of the American Philosophical Society and met Thomas Jefferson there on March 10th, 1797 when Jefferson presided at the session.[1]

Joseph Priestley House: Home to Rev. Dr. Joseph Priestley

As reflected in Priestley's letters, Mary Priestley is credited by her husband with planning the mansion situated just outside the village on land along the banks of the Susquehanna River.

On August 26th, 1794, Mary wrote to family friend William Vaughn in London that she and Joseph were planning to build a house for permanent occupancy: "We have taken some ground and are now in treaty to have a frame house built upon it to live in", and promised to send the house plans to William: "At some future period I will send you a plan of our house with the extent of our premises and the view from it."[2]

Joseph Priestley hired a carpenter, Mr. Jones, to lead the construction work and building began. In October 1794, Priestley wrote to his friend, Benjamin Rush, a Philadelphia physician, asking Rush for assistance to find skilled mechanics capable of building a suitable house who could be sent on to Northumberland to assist Mr. Jones. Within a few weeks, Priestley wrote to his brother-in-law, John Wilkinson, "As houses are not to be got here, I have begun to build one, but the expense will be great on account of the enormously high price of labour, and the scarcity of workmen of all kinds. If this circumstance was properly known in England, we should draw many thousands of Carpenters, Smiths, Masons, and handicraftsmen of all sorts ... Even now I give near three shillings per day for mere labour ..."[3]

One year later, in October 1795, Priestley made a decision to use wood, a readily available Pennsylvania building material, for the frame and exterior clapboards of the permanent house since local weather conditions at the time delayed the making of bricks – his preferred building material.

While house planning and construction were underway, Joseph and Mary lived in a number of other locations, and for at least the last six months of 1794, Priestley's books and laboratory apparatus were stored in his son's garret. In August 1794, Mary wrote, "We have a good house the best in the place, but we can only have it till May".[4] The house Mary referred to in her letter may have been that of Josiah Haines of Philadelphia, as mentioned by Mr. William Bakewell, Haines's gardener and house steward, who arrived in Northumberland in the spring of 1795 and found Priestley was a 'resident in [Josiah] Haines's house while his own was fitting up'[5]

Bakewell's reference to the fitting up of Priestley's 'own' house at this time probably referred to the temporary house Priestley was in process of securing. In a letter written to George Humphries in October 1794, Priestley indicated he had entered into an agreement for a house with a Mr. Jenkins.[6]

Thomas Russell, son of William Russell, knew the Priestleys from their old neighbourhood in Birmingham. Thomas visited Northumberland and described Joseph Priestley's small house as "a frame house ... fitted up with all that neatness for which his wife is well known, but still it is a mere hut in comparison with the one they lived in formerly."[7]

The property Priestley bought consisted of two lots #55 and #56 with a house situated along Orange Street between Water Street and North Way at a cost of £300. The location was just across North Way from the four lots where the Priestley's frame house would be under construction.[8]

From May 1795 until they moved into the temporary Lot #55 house later that same year, Joseph and Mary probably lived with their eldest son, Joseph, and his wife, Elizabeth, in their brick house nearby.

From the temporary house, Mary could watch the progress being made on the construction of her new house and continue with plans for furnishing and decorating the home upon its completion. These happy plans were not to be. Mary's long-standing health problems increased when her youngest son, Henry, died of a fever in December 1795. She had increasing episodes of blood-spitting and was carefully nursed by her daughters-in-law and Mr. Bakewell's wife, but died on September 7th, 1796 while her house was still under construction.

After Mary's death, Joseph Priestley, in mourning, returned to live in his son Joseph's home. Despite set backs due to weather-related problems, difficulties with finding skilled carpenters and inflated costs of construction, Priestley's frame house neared completion late in 1797. By September of that year, the laboratory was finished and Priestley was able to set up his apparatus and resume experimenting for the greater part of the day.[9]

Priestley took up occupancy of his large, new home in 1798, possibly in the spring, and invited his son Joseph and his family to live with him. While Joseph Jr. was much involved with land dealings and financial matters, his wife, Elizabeth, took her deceased mother-in-law's place as mistress of the intergenerational household. Elizabeth brought her two children, Joseph and Elizabeth, to live in her father-in-law's house. She cared for her children, saw to the day-to-day affairs of the household, and supervised the work of the family's three servants – a hired man, hired girl and little bound [indentured] girl.[10]

Priestley worked from time to time in his Northumberland laboratory where he isolated carbon monoxide in 1799. He continued as a prolific writer of pamphlets and books containing his theological and political thoughts and opinions, and philosophical research results. Priestley seemed to enjoy living in the new house but wrote in 1800, "Indeed, I am incapable of managing a house myself, and always was. While my wife was living, I said I was only a lodger in her house, as I am now in my son's ...", an indication that Priestley had turned over responsibility for running the household to the younger generation.[11]

Priestley died in 1804 in the library of his Northumberland home in the company of his son, Joseph, his daughter-in-law, Elizabeth, his grandchildren and his friend, Thomas Cooper.

After his father's death, Joseph Priestley Jr. wrote about his father and his house: "Towards the end of 1797 and not before, his library and laboratory were finished. None but men devoted to literature can imagine the pleasure he derived from being able to renew his experiments with every possible convenience, and from having his books once more arranged. His house was situated in a garden commanding a prospect equal, if not superior, to any on the river Susquehanna, so justly celebrated for the picturesque views its banks afford."[12]

Joseph Priestley House: Home of Joseph Priestley Jr.

After the death of Rev. Dr. Joseph Priestley, his son and namesake Joseph Priestley Jr., who became a naturalized citizen in 1798, inherited his father's house and its contents, including the library holdings, scientific apparatus, and interest on investments made in French funds. He continued to reside in the Northumberland home with his wife and children for some years.

Joseph was much involved with his numerous properties and land holdings. He served his community as Justice of the Peace for Point Township and Northumberland, having been appointed to the post by Governor Thomas

McKean on April 2nd, 1803.[13] In 1808, he was appointed as a commissioner of the Centre Turnpike Company which was incorporated by an act of the Pennsylvania Legislature in 1805 to construct a road from Sunbury to Reading. Joseph also served as a commissioner on the Northumberland Bridge Company formed in 1809 to build bridges from Northumberland to Shamokin Island and thence to Sunbury.[14]

Pennsylvania's Governor Snyder accepted Priestley's resignation as Justice of the Peace on March 5th, 1811. Shortly thereafter, Joseph made a valuation of all his property. In addition to his father's house, he recorded ownership of numerous properties throughout Northumberland and Point Township including farms and nurseries, land in Lycoming County, and livestock. He held a number of mortgages, owned shares of stock in the Turnpike and Bridge companies, and enumerated debts owed him by private individuals.[15]

Later that year, Joseph departed Northumberland with his family for Philadelphia and from there returned to England early in 1812 where he remained. Acting through his attorneys, Thomas Cooper and John Vaughan, Joseph sold the Priestley house and its adjoining lots on May 13, 1815 to Seth Chapman, Esquire. From that point until 1919, numerous families bought, rented, lived in and sold the house.[16]

Later Occupants of the Priestley House

Seth Chapman lived in the Priestley House, first as a tenant from 1811 until 1815, and then as owner of the property having purchased it for $6240 from Joseph Priestley Jr. Chapman was a Bucks County lawyer with 20 years of experience when he was commissioned in 1811 as president judge of the Eighth Judicial District composed of Northumberland, Luzerne and Lycoming counties. As the law required a judge to reside within the limits of his district, Chapman moved to Northumberland. According to census records from 1820 and 1830, Judge Chapman, his family and servants, including in 1820 a free coloured female age 16–26,[17] and in 1830 a free coloured male age 10–23[18] occupied the premises. Chapman's career ended in retirement in 1833 after an unsuccessful impeachment trial based on allegations of favouritism and undue partiality. He died in 1835 and was buried in Northumberland's Riverview Cemetery.[19]

It was during this time that canal construction along the West and North branches of the Susquehanna River began. A portion of the canal along the North branch was dug through the lower portion of the former Priestley property. In 1830, the canals opened for transport and the first two packet boats for passengers, the *George Denison* and the *Gertrude*, operated between Northumberland and Wilkes-Barre beginning in 1835.[20]

In 1844, nearly ten years after Chapman's death, the Orphan's Court of Northumberland County ordered the sale of Chapman's real estate. It was purchased by his two sons-in-law. John A. Sterrett and his wife, Sarah Chapman, purchased three-fifths interest and Charles H. Kay, husband of Jane

Chapman, bought the remaining two-fifths. In 1847, Sterrett sold his interest to Charles Kay for $7500. Charles Kay died in 1851, but his widow Jane kept the property until 1865. During this period, there were two tenants in successive occupation. The first tenant was Charles Kay's father, the Reverend James Kay, an Englishman called to lead the Unitarian congregation in Northumberland from 1822 until his death in 1847.[21] Later, John J. Reimensnyder, a Lutheran pastor and farmer, occupied the house with his family until 1860.[22]

Jane Chapman Kay sold the house and lots fourteen years later in 1865 to Henry R. Campbell of Vermont for $2775. For three years, Campbell stayed in the house while on business trips. He was witness to the 1866 explosion of a locomotive standing near the house causing damage to a corner of the house in the vicinity of the kitchen.

In 1868, Campbell and his wife, Sydney, sold the house and lots for $5679.33 to Florence M. Bingham of Philadelphia. While Bingham owned the property, the house was occupied by tenants.

In 1882, the heirs of Florence Bingham conveyed the property to Thomas Hugh Johnson of Northumberland for $2000. It is believed the house remained a rental property.

In 1888, Johnson and his wife Ann sold the property to Kate Scott of Northumberland for $3000. For some years the Scott family occupied the house and may have sublet it in 1910–11 as a boarding house for labourers. At about the same time, a branch railway line was cut through the lawn where the canal had broken the slope to the river.

By 1919, the Priestley House had become shabby and in need of repairs. The landscape was overgrown and it was rumoured the house was haunted and would be torn down for expansion of the railway freight yards in Northumberland. The heirs of Kate Scott conducted a public sale of the house and lots at which Dr. George Gilbert "Swampy" Pond, Professor of Chemistry and Dean of the School of Natural Sciences at Pennsylvania State College, now Pennsylvania State University, purchased the property for $6000 with the intention of moving the house about 80 miles west to the State College campus.[23]

Joseph Priestley House: "Swampy" Pond's Dream & the Penn State College Days

On January 13th, 1920, Professor Pond wrote the following letter:

> Dear Boys,
>
> ... The Priestley House is ours! It will come to the Penn State Campus. It will be restored here with its original frame, and enough old material to really be the same old mansion, but it will be modern and useful. It will be endowed with $10000 in Liberty Bonds, so that its material upkeep will be assured.
>
> I now believe it is most likely to be used as an administration

building as regards the ground floor; quarters for bachelor instructors on the second floor; and possibly rooms for society meetings on the third. What an inspiration to those who work and live under that roof!

Priestley's Laboratory is, of course, the most precious part of it all. That must be preserved as an especially valued relic, in the form of a museum of some sort. Your contributions will perhaps determine the degree of perfection with which that feature can be elaborated. ...

Your ever affectionate,
OLD SWAMP[24]

Professor Pond died before actually completing the intended conveyance of the title to Pennsylvania State College to operate the house as a chemistry shrine. At the time of Dr. Pond's death on May 20th, 1920, the George Gilbert Pond Memorial committee was founded. While the house remained the legal property of Pond's heirs until 1932, the college's committee assumed responsibility for the property, established a fund to preserve and maintain the house on its original site, and removed most traces of previous occupants.

In 1926, the college built a separate fireproof brick museum building on the property as a memorial to Professor Pond. Designed by A. Lawrence Kocher, a member of the Penn State faculty, the building housed a display of Priestley's books, manuscripts, scientific apparatus and other memorabilia offered by family descendants, private collectors and the Smithsonian Institution. The artefacts, some in poor condition, were eventually removed from the unheated building after a number of years due to dampness and mildew problems and returned to their lenders or transferred to the Penn State library.

During this time, Mr. John L. Bingaman was engaged as caretaker and lived with his family in the house until his death in 1934. After Mr. Bingaman's death, Mr. Russell A. Clugston lived in the house as custodian until April 1937.[25]

Subsequently, George and Suzanne Neff of Sunbury, leased the house and occupied the second floor in return for keeping the house and grounds in good repair. The Neffs expected to completely restore the Priestley House based on their study of colonial Virginia mansions. They were, in essence, the first curators of the property and conducted tours of the first floor of the house.[26]

In a memo written May 5th, 1954, Kenneth R. Erfft, Associate Comptroller at Penn State, noted that between 1943 and 1949, the college spent a total of $4565.79 in routine maintenance of the Priestley House and faced a minimum of $3200 for additional, urgently needed improvements including a new stoker for the residence, painting of buildings, rebuilding fences and placing a new roof on the Museum building. Erfft went on to recommend a decision be made as to the extent to which the University planned to continue its support of the Priestley House.[27] In 1955, the officers of the University transferred the deed for the Priestley property to the Borough of Northumberland for $1.

Joseph Priestley House: 1955 to 1990

The Northumberland Kiwanis Club, a civic organization, encouraged the borough's acquisition of the historic property. Local residents served on the Priestley Home Association, a committee that represented the borough council and interested civic groups.[28] The Association hired caretakers and tended to the maintenance and preservation of the house until 1961. At that time, the borough conveyed the property to the Commonwealth of Pennsylvania for $1.[29] Since then, the Pennsylvania Historical and Museum Commission has managed, supervised, maintained and improved the property.

Under the Commission's administration, the Priestley House was designated as a National Historic Landmark by the National Parks Service, U. S. Department of the Interior, in 1965 for its historic and architectural significance as an early example of Federal architecture in the rural setting where Dr. Priestley lived and worked. Subsequently the property qualified for and was named to the National Register of Historic Places. The Commission began restoration of the property in 1968 and reopened the Joseph Priestley House for public visitation in October, 1970.[30]

In 1989, the entire Historic District of Northumberland, including the Joseph Priestley House, was named to the National Register of Historic Places based on significant events in local history and the wide variety of surviving architectural building styles.[31] These special designations will insure that the Priestley House, as a unique element of Pennsylvania's history, will continue to be preserved for the benefit of future generations.

Joseph Priestley House: Interpretation for the Future

Since 1990, the Priestley House historic site has progressed in research, preservation, interpretation and public programs under the direction and leadership of professional museum staff. Re-interpretation of the interior spaces of the house was completed in 1994 based on the 1800 architectural drawing of the house and landscape by T. Lambourne. Priestley's library, drawing room and dining room spaces were identified and furnished with period furniture and many artefacts with Priestley provenance, including books, scientific apparatus, clocks and personal items.

On August 1st, 1994, the American Chemical Society designated the Priestley House a National Historic Chemical Landmark honoring the site of Priestley's American laboratory where he first isolated carbon monoxide as "... probably the best in the country at that time".

A carriage barn, reconstructed in 2000 as depicted on the 1800 architect's drawing, now houses a modern visitor centre with orientation exhibit and video, gift shop and accessible restrooms. Guided tours of the house are available for individuals, groups and school students. Research opportunities are available by appointment.

Current preservation efforts focus on the historic laboratory, a unique example of an early, well-equipped scientific research space in America, even more unique in its highly rural Northumberland location. Recent archaeological excavation conducted in the laboratory revealed brick supports for furnaces in two corners of the room, while architectural examination of the laboratory found evidence of a fume hood that provided vital air exchange integral to Priestley's experimental work in pneumatic chemistry. Artefacts on display in the laboratory are largely reproductions of Priestley's original glassware, apparatus and machinery now in the collections of Dickinson College and The Franklin Institute in Pennsylvania, and at the Smithsonian's National Museum of American History, Washington, D. C., as well as other institutions, in accordance with the wishes of Joseph Priestley's family members and friends.

Research on the laboratory is on-going with a goal of interpreting the laboratory as accurately as possible to depict Priestley's scientific work undertaken while in Northumberland.

Research partnerships with individuals knowledgeable in historic American and British laboratories of the eighteenth century are sought. While the research process continues, visitors can access the laboratory via a temporary floor that allows visitors to see the archaeological remains of the furnaces. Staff guides and docents will inform visitors about research discoveries as the project moves forward.

Outreach programming and alliance with local and regional communities, universities and historical organizations, and the American Chemical Society has been made possible with assistance provided by the Friends of Joseph Priestley House, a not-for-profit volunteer organization that supports educational programs and raises funds for various initiatives related to Priestley House. In October 2001, the Friends cooperated with Penn State University's Chemistry Department to plan, organize and present a public program recognizing the 125th Anniversary of the founding of the American Chemical Society, a professional membership group with connections to Priestley House since 1874 when chemists, some of whom later organized the Society, visited Priestley House to commemorate the Centennial of Priestley's discovery of oxygen.

The on-going commitment of the Pennsylvania Historical and Museum Commission to preserve and interpret the Joseph Priestley House with the assistance of the Friends of Joseph Priestley House remains vital and, in concert with public interest and advocacy, will enable people who visit Joseph Priestley's American home to learn about his scientific achievements and understand his impact on American education, government and theology.

NOTES

1. James J. Bohning. "Joseph Priestley and the American Philosophical Society" compiled from "Early Proceedings of the American Philosophical Society 1744 to 1838" {Philadelphia 1884). June 2003.
2. Mary Priestley. Copy of letter to William Vaughn, August 26th, 1794. National Endowment for the Humanities grant research file folder 39–48, Joseph Priestley House.
3. Joseph Priestley. Copy of letter to John Wilkinson, November 12th, 1794. NEH grant research file folder pp. 39–48, Joseph Priestley House.
4. Mary Priestley. Copy of letter to William Vaughn, August 26th, 1794.
5. William Bakewell, "Some Particulars of Dr. Priestley's Residence at Northumberland, America," *The Monthly Repository of Theology and General Literature*, vol. 1, No. VIII, (London, August 1806): p. 394.
6. Letter from Joseph Priestley to Mr. George Humpries, October 15th, 1794. Lyman C. Newell, "One of Priestley's First Letters Written from Northumberland, PA," *Journal of Chemical Education*, vol. 4, no. 2 (February 1927) : pp. 173–74.
7. S. H. Heyes, *The Russells of Birmingham*,. (London: Allen & Company, Ltd., 1911), pp. 197–98.
8. Robert G. Christ, Report #12 to Tomas Spiers, August 25th, 1981. Christ Research File 6, Joseph Priestley House.
9. Joseph Priestley, *Memoirs of Dr. Joseph Priestley To The Year 1795 with a Continuation by His Son, Joseph Priestley*. (London: J. Johnson, 1806), pp. 194–95.
10. Leslie Patrick, "Priestley and the Inhabitants of Northumberland: Slaves and Servants, Friends and Acquaintances," Final report for Raising Our Sites: Community Histories of Pennsylvania, October 2000.
11. Joseph Priestley, Letter to Mr. Belsham, June 5th, 1800. J. T. Rutt, ed., *The Theological and Miscellaneous Works of Joseph Priestley* (London 1817–1831), Parts 1 and 2: Life and Correspondence, II, pp. 437–8.
12. Joseph Priestley, *Memoirs*, pp. 194–95.
13. Robert G. Christ, research file on Joseph Priestley Jr. NEH grant research file folder 7, Joseph Priestley House.
14. Herbert C. Bell, *History of Northumberland County*. (Chicago: Brown, Runk & Co., 1891), pp. 305–6, 528–30.
15. Christ research file on Joseph Priestley Jr., NEH grant research file folder 7.
16. Tomas Spiers and Associates, "Historic Structure Report for Joseph Priestley House," (1981).
17. Robert G. Christ, Report #3 to Tomas Spiers, 1981.
18. U. S. Census research by Katherine Hastings, 2004.
19. Ref. 14, pp. 223–26.
20. Ref. 14, pp. 309–11.
21. Ref. 14, p. 539.
22. Hastings, 1860 U. S. Census research, 2004.

23. Ref. 16, Part 2, pp. 15–21.

24. Charles G. Murphy, "The Reverend Dr. Joseph Priestley's American Home, Northumberland, PA," undergraduate research paper in Historic Preservation (1977), Robert G. Christ research files, Joseph Priestley House.

25. Copies of Penn State College correspondence 1926–1955, site research files, Joseph Priestley House.

26. Daily Item, Sunbury, PA, "Priestley Home Revelation After Restorative Work," (June 7th, 1937), Penn State College correspondence, site research files, Joseph Priestley House.

27. Copy of memo K. R. Erfft to S. K. Hostetter, May 5th, 1954, Penn State College correspondence, site research files, Joseph Priestley House.

28. Daily Item, Sunbury, PA, "Priestley Story Recalled as Borough Acquires Shrine," (December 14th, 1955), Lester Kieft research files, Joseph Priestley House.

29. Unknown newspaper article, "Governor Signs Bill for State to Take Over Priestley Home," (pre- July 1, 1960), press release files 1950–89, Joseph Priestley House.

30. Ref. 16, Part 2, pp. 37–38.

31. Reading Eagle, Reading, PA, "State Cites Historic Northumberland Area," (January 8th, 1989), press release files 1950–89, Joseph Priestley House.

SUPPLEMENT – PICTURES FROM JOSEPH PRIESTLEY HOUSE

Priestley's Northumberland house – the study

Priestley's Northumberland house – the laboratory

Priestley's telescope, glasses and logarithmic calculating 'bones'

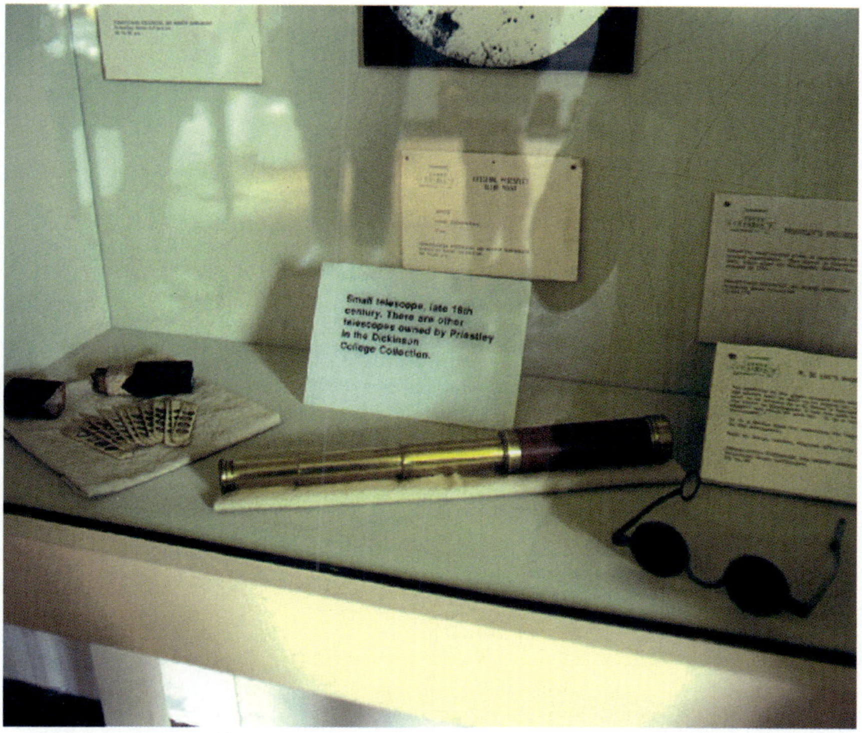

Replica chemical/gas apparatus

Replica electric spark/gas experimental apparatus

DISCOVERING AIRS OF MANY DIFFERENT KINDS

Ronald C. Blatchley

Ronald C.Blatchley

The concept of "air" goes back to the ancient Greeks who counted it among their four elements. They thought of it more as a *quality* than a physical substance and the idea persisted into the 17th Century.

Jan van Helmont (1577–1644) studied the properties of airs produced by fermenting grapes, burning charcoal, or pouring vinegar on seashells and noted their similarity. His gases were impure and so did not always possess the same characteristics but he was more interested in physical properties than chemical properties and was unconcerned with the lack of reproducibility in his experiments. Air was useful for fanning a flame or for absorbing products of combustion. But not much else. He invented the word "gas" (derived from the Greek "chaos") to describe such substances. The derivation may have been influenced by his observation that attempts to make airs frequently resulted in broken glassware. Gases beget chaos. The new word was used sporadically on the continent but the influence of Robert Boyle (1627–1691) kept "air" in our English vocabulary until the 19th Century.

Boyle and his associate Robert Hooke (1635–1703) were, like van Helmont, preoccupied with the physics of airs more than the chemistry. They understood that part of the air was necessary for combustion and that otherwise combustible substances would not burn in a vacuum. Their chemical efforts were hobbled because they lacked convenient ways to generate, collect, and experiment with gases.

Stephen Hales (1677–1761) provided the missing technology when he separated the generating vessel from the receiving vessel with the invention of the pneumatic trough. He extracted airs from a variety of biological substances: wood, seeds, sugar, urine, etc. His method was to place the substance in the closed end of a bent gun barrel and then heat that end in a fire. As airs issued from the other end of the barrel, he bubbled them into a receiving vessel. He was chiefly interested in the *volume* of air produced and paid scant attention to chemical properties. It is very likely that he produced several common gases in his experiments but failed to observe and describe their identifying characteristics. He also failed to recognize that some of his gases were soluble in water, rendering much of his quantitative date meaningless. It remained for our own Joseph Priestley to solve the problem by collecting gases over mercury rather than water.

The discovery of airs really began with Joseph Black's (1728–1799) 1754 doctoral dissertation in which he described the reaction of acids with *magnesia alba* (magnesium carbonate). Black collected the air (carbon dioxide) produced by the reaction and described its properties. He also observed that the same air could be produced by heating *magnesia alba* and that the air could be "fixed" into a solid by reaction with lime. Black's innovative experiments with fixed air made him the first pneumatic chemist; his predecessors were at best, pneumatic physicists.

Henry Cavendish (1731–1810) was – for all his eccentricities – one of the most meticulous experimenters of all time. He prepared *inflammable air* from the reaction of several metals with a variety of acids and reported his findings in a series of papers to the Royal Society in 1766. He dried the airs, compensated for temperature and pressure differences and showed the air to be chemically and physically identical in each case.

Suddenly it was clear that many *factitious airs* awaited discovery and many professional and amateur chemists went looking for them. None was more successful than Joseph Priestley who is credited with the discovery of at least eight kinds of air, the most famous of which was *dephlogisticated air*. Some controversy continues even today concerning the true discoverer of oxygen: Scheele, Lavoisier, or Priestley. But all agree that Priestley was first to prepare and collect pure oxygen, describe its chemical, physical, and biological properties ... and publish his findings. Others – even as early as Boyle – may have preceded him in some parts of the work but it was Priestley who did it all.

Using that same set of criteria, Priestley also discovered nitric oxide, nitrogen dioxide, nitrous oxide, carbon monoxide, sulfur dioxide, ammonia, and silicon tetrafluoride. He studied – but probably did not "discover" – hydrogen chloride, sulfur trioxide, chlorine, phosphine, hydrogen sulfide, and nitrogen. Except for the organics and the Noble Gases, he did not leave much for anybody else to do.

His work was marked by inventive genius and skilful use of apparatus but was also hampered by his passionate embrace of phlogiston theory and an avoidance of measurement and mathematics.

Priestley has sometimes been called "The Father of Modern Chemistry" but the antiphlogistonists, however, would to this day dispute that honour in favour of Lavoisier.

PRIESTLEY COMES TO OAKWELL

On the evening of 7th February 2004, Friends of Oakwell Hall hosted a special Candlelight Historic Supper in the delightful Hall in honour of our special guests, Friends from Joseph Priestley House Museum in Pennsylvania. Dressed in costume, many of the Oakwell Friends re-enacted 17th Century activities, including the preparation of the "gran salat" in the kitchen, quill-pen writing, pomander making, potpourri bags, and "godseyes". Guests had a guided tour of the Hall followed by a XVIIc buffet supper

Joan and Beatrice preparing the "Gran Salat"

On Sunday afternoon of 8th February, the Birstall festivities were concluded with the presentation by the Priestley Society and staff at Oakwell Hall of the Science Spectacular **'OXYGENIUS'**: a celebration for all the family. The aim was to illustrate the many facets of Priestley's scientific work and related subjects; oxygen and airs, photosynthesis, electricity, and of course a good supply of cans of Coca-Cola, free to those children who remembered the correct answer to the question "Who invented fizzy drinks?"

'Joseph Priestley' (alias Ron Blatchley) performing his experiments

The highlight of the OXYGENIUS educational activities was Joseph Priestley himself (Ronald Blatchley in costume), who entertained as he conducted experiments in the spirit of Priestley. As he talked the audience through many of Priestley's discoveries, Ron joked "The mice that Priestley experimented on without air didn't die... he gave *mouth-to-mouse* resuscitation!".

(Friends of Oakwell Hall: Newsletter No. 42, 2004, extract from article by Joan Edwards: www.oakwellhall.f9.co.uk)

Priestley's West Riding Heritage

DR PRIESTLEY'S WEST RIDING

John Goodchild
The John Goodchild Collection and Archives[1]
(An independent Local History Study Centre for the Central West Yorkshire)

John Goodchild

It is a curious coincidence, that we meet this evening on 4th July, United States Independence Day, for the United States was seen in Priestley's day among people of his background as the land of religious liberty, and in his adulthood as a land of liberal political experiment, as well as one providing – and to an extent in association – major markets for West Riding cloth. It was viewed as a land of Englishmen across the Atlantic and as a land of liberty and independence and potential, and there was much sympathy for their revolution from the 1770s and independence from the 1780s: the still standing Triumphal Arch at Aberford is but one surviving sign of local liberal enthusiasm for their cause, erected as it was to commemorate the successes of the erstwhile revolting colonists in northern America.

Ideas and enthusiasms were shared between Yorkshiremen and those in the distant colonies, as well as trade; even numbers of boys were sent across to England to be educated at Yorkshire schools. Priestley himself was to migrate to the new United States, subsequent to his becoming a kind of Arthur Scargill hate figure in England, through the picture of him projected by the English government and media.

England, and Yorkshire as much as any part of this country, was changing markedly too. Priestley's West Riding of Yorkshire was just as much a land of innovation, in vibrant enterprise, in political theory and to a degree in political practice, as it was in religion, in manufacturing, in commerce, in science, in housing, in the growth of a new and powerful middle class, in the provision of social and educational facilities, and local amenities.

Much of the change of this period, of Priestley's lifetime from 1733 to 1804, is reflected in the pages of the press of the time: a newspaper had been established at Leeds in 1718, one in York in 1719 and others subsequently at Sheffield, Halifax and Wakefield – sometimes rival newspapers were published, as with

Felkirk Parish Church, Shafton Village 1732

Felkirk Church, Shafton Village, 1732 – a painting by modern artist Stephen Barlow. Shafton, about 6 miles from Wakefield, was the birthplace of Joseph Priestley's mother, Mary Swift, in 1706. The church is where she married Joseph Priestley's father, Jonas, in 1732. Joseph spent his formative years with his maternal grandparents at their farm in Shafton until his mother's death in 1739.

the Leeds paper which is now the Yorkshire Post, established in 1754. In their obvious and in their unwitting testimonies, from news items, from advertisements, from obituary notices, they tell much of the concerns and manner of life of Priestley's time here in the West Riding, and *inter alia* they refer to Priestley himself, while other and more specific source materials have survived in considerable bulk to illustrate local life too.

Population was growing rapidly: relatively reliable overall figures are only available from the first national census of 1801, but in some areas earlier, unofficial censuses were taken, which with parish and nonconformist registers, also illustrate this growth. Inclosure was changing the face of town and countryside by its dividing into modern fields (with potential now for development), of large areas of hitherto common land, while many of the towns of the West Riding which came in Priestley's period to be described as booming, were advantageously owned by small freeholders – many of them political Whigs – and only Huddersfield and Sheffield among them were largely owned by single estate owners, the baronets Ramsden and the Dukes of Norfolk respectively.

Industry, commerce and trade were all much encouraged by new transport improvements. Improved roads came with toll roads or turnpikes, set up in the West Riding in increasing numbers from 1735, while old road routes were soon afterwards being improved in their hills, bends, even mileages, and entirely new road routes being built by the end of the 18th century, not the least of the improvements being effected by the almost unbelievable but very real abilities of Blind Jack of Knaresborough – John Metcalfe, the blind road engineer. One such turnpike ran through Priestley's home village of Heckmondwike.

The cheaper-to-use inland waterway system of the West Riding – a totally inland county in itself – was introduced with the opening of the Aire and

Calder Navigation to Leeds in 1700, with a branch to Wakefield immediately subsequently, which provided cheaper transport for the importation of raw wool, the export of cloth, the importation of timber and the transit of other raw materials and manufactured goods; but especially did it open up the hitherto landlocked West Riding coalfields to enormous new markets.

The company's investors were to receive dividends, which rose to some 20% in the mid-18th century and 100% by its end. The construction of further independent but connecting navigable waterways and later artificial canals greatly augmented the ease and cost of carrying, and these waterways can be characterised as the motorway equivalents of their time. In the year of Priestley's death in 1804, the first of what were ultimately three trans-Pennine canals was opened between the West Riding and the industrial North West.

Railways too were not unknown in Priestley's day, even though merely feeders to waterways or nearby large towns demanding coal, iron or lime: the first British Railway Act of Parliament was passed in 1758 to enable the construction of part of a railway near Leeds, and in 1798 was opened in the West Riding the world's first public railway, built for the world's earliest known railway company.

But what did these improved transport facilities assist? Primarily in the West Riding they gave birth to a great coal industry, and vastly assisted the woollen textile industries, one largely in woollen and worsteds, but also in linen and even cottons. To facilitate the sales of woollens and worsteds as between handloom weaver and cloth merchant, great cloth-selling halls were built in Priestley's lifetime, at Leeds (two of them), at Halifax and Huddersfield, at Penistone and Wakefield, even at Gomersal. One of these great Leeds cloth halls was opened in 1758, in the fields adjoining Mill Hill Chapel, now City Square.

But technology was, as ever advancing too, and from the 1780s new and increasingly steam-engine-powered mills were being built, their machinery producing yarns, while West Riding woollen cloths were being exported internationally, including cloth for the imperial armies of Catherine the Great in Russia. The multi-process cloth factory, bringing in raw wool and sending finished cloth through its gates, came first on a large scale, with Benjamin Gott and his partner's mill in Leeds in the 1790s.

The iron industry became here from the 1780s of increasing, even national significance, with associated engineering and steam engine building, while widespread agricultural improvement was made, including increased use of new technology there too.

There were many topographers and commentators who described the West Riding of their own day in some detail in the late 18th century, and for us they provide useful insights. But they say little of the developing infrastructure which so improved life at that period: of the friendly societies by which the less well off insured themselves by providing their own funds for unemployment, illness, death and sometimes pensions; of the entertainment facilities of pubs, clubs, theatres, races or hunts; of the social provision of hospitals, libraries, of Masonic lodges, of early cooperative and building societies. There was a widening of the powers and numbers of magistrates, a provision of water supplies, of town drainage, of policing, of street paving; a growth in suburbs, of both working and middle class natures. The poor laws were being implemented in a necessarily numerically wider but often liberal manner; formal education for the working class was increasing in its extent; Sunday Schools were established from the 1780s, teaching the 3Rs; new bridges were being built metaphorically as well as really. Society and the economy, not to mention the landscape, were changing radically.

The concerns of the mind were changing too: books and newspapers and education for the poorer were newly both readily and cheaply available, while sons of the middle class went in larger numbers to Oxbridge, or to the Scottish universities, or to the nonconformist academies.

Contested county elections may have been few for those males who possessed the 40 shillings a year annual value freehold votes; in any case, costs kept down contested elections, and after that of 1741, when Priestley was a boy, there was not another until 1807, while the very few Parliamentary boroughs in the West Riding were largely controlled by major landowners – but interest in politics was great, and as land was divided and the 40 shillings voter qualification decreased in real value, the number of county voters increased: in 1741, 14 had voted from Priestley's Heckmondwike: in 1807 there were 174.

Tory rampaging mobs were indeed not unknown, but in the West Riding the Glorious Revolution of 1688 had been largely supported and there was strong opposition to the Stuart rebellion of 1745, with much subsequent support, largely visible in the newly expanding middle class, for both the American revolution and the French revolution in its earlier stages: one middle class boy in Wakefield was given, early in the 1790s, the forenames of "Alfred Washington Mirabeau", those of the fathers of liberty as then seen. In science, an informal group of cognoscenti, a kind of northern Lunar Society, met at the opulent rectory at Thornhill, and religion was very much literally at the heart of society. Roman Catholicism was in a parlous state, but dissent in the forms of the English Presbyterians or Unitarians, the Independents or Congregationalists, the Baptists, the Quakers, flourished in the industrialised areas of the West Riding, when the Church of England was largely, though by no means wholly, as in Heckmondwike quiescent. From the 1730s came the Methodists, whose initially condemned enthusiasm ultimately transferred to the Old Dissent and the Church – even perhaps in a degree to the Romanists. The Methodists' essential concern with the religion of the heart as opposed to that of heart and head combined, was to receive Priestley's admonitions, but the Methodists grew in number, although very largely claiming the allegiance of the upper working class in urbanised areas – and were to split into sects after John Wesley's death in 1791.

The milieu in which Priestley grew up was one typical of one part of middle class of the manufacturing parts of the West Riding of that time: of the comfortably off, though by no means wealthy, those of mildly nonconformist sympathies in religion and politics, well able to read and understand much of the latest philosophical, political, social and theological thrusts of time. Importantly, such matters were openly discussed in Priestley's aunt's house, his own home, and discussed from all viewpoints, and this was a major influence of Priestley's own development, as he writes in his Memoirs.

When Priestley left home at 17, in 1750, the West Riding was about to embark upon a period of accelerated industrialisation, of huge boom and change in its appearance, its economy, in its political and religious awarenesses and expressed sympathies, but his experiences at home were essentially those of incipient change. He was later to witness immense change and to a degree to take some part in it himself as a result of his own scientific, his political, his educational and his religious thinking and exploration, and his propagandist abilities. I would argue that Priestley was initially and basically necessarily shaped by the world he experienced here in the West Riding; how he went on to take an important part in the further reshaping of the West Riding and the world is Priestley's life story.

In so many ways the West Riding of Priestley's lifetime was similar to that of our own: for it and ours were and are times of radical change. Priestley's peers, like ours, lived in a time of international wars, in one of far reaching economic and technological change, of new economic and employment circumstances, of new ideas in science, art, architecture and literature, one of political machinations; in one of far reaching educational and health national and international reform, one of transport difficulties, of changes in the use of the countryside, of vast social and political inequalities, of increasing religious fundamentalism, of political dissatisfaction, of interest indeed in moral and spiritual questions, and the parallels continue. But Priestley's time coped with its troubles, as we will no doubt muddle through ours. My concern has essentially been to point out the constant nature of human concerns as times change.

NOTES

1. THE JOHN GOODCHILD COLLECTION AND ARCHIVES is privately owned and funded by John Goodchild. John formed the study centre in 1994; it now contains thousands of documents and maps, etc., charting local history, and is visited by around two thousand historians from across the world every year, these including historians from as far away as Australia, USA and Taiwan. John has pursued his research and publication of local Wakefield history, and has written in excess of two hundred books and articles. He has accumulated the collection over his lifetime, and access to the public is free of charge; the whole vast collection is left at John's death to the local authority. At present, it is situated in the basement of the Drury Lane Library in Wakefield.

SECTION 5

England Once More

A PLAY IN TWO PARTS BY ADAM STRICKSON

The play 'England Once More' begins with a depiction of some of the key events in Joseph Priestley's life from 1791. It was written by Adam Strickson, working with sixth formers from Woodkirk High Specialist Science School. They wanted to portray as much as they could of the history of the times and spirit of the man in a short drama. In the final scene Priestley returns to the Mill Hill Chapel in 2004 and, in conversation with three teenagers, something of his beliefs and the phlogiston theory are made plain. Woodkirk High School is only a few miles from Priestley's birthplace. In Autumn 2003, it was looking for a novel project to mark its award of specialist status. Approached by Education Leeds, who wanted to celebrate Priestley's bicentenary, the decision was taken to commission a play. The result was researched, acted and partly written by the students. They also wrote and performed the background music and song. After live performances in school, Mill Hill Chapel and Oakwell Hall, a studio performance was filmed.

Virginia Stokeley, Director of Specialist Science,
Woodkirk School, West Yorkshire

Adam Strickson
(Playright)

Virginia Stokeley
(Producer)

ENGLAND ONCE MORE,
OR, THE LIFE, DEATH AND UNEXPECTED RETURN OF JOSEPH PRIESTLEY

By Adam Strickson
With Sixth form students from Woodkirk High Specialist Science School

Characters

Joseph Priestley aged 58–71.
Mary, his wife, a little younger.
Martha Russell, a family friend in her late 20s.
Fina, Lucy and Jenny: three university friends living in Leeds. Fina is studying a science-based subject.
Joseph's three grandchildren (the same actors as the three friends).
Voices of protesters at Joseph's sermons in Birmingham and in present day Leeds.

Voices of rioters in Birmingham, wrecking Joseph's house.
Timothy, Joseph's brother; his mother and his Aunt Sarah (pre-recorded voices).
Singers (who could be ragged sailors) on board The Sansom.

Part 1: The last years

Before the play, a projection of the cartoon of Priestley as Doctor Phlogiston

Scene 1. Birmingham, 1791

A projection of the pulpit at Mill Hill Chapel/the actual pulpit at the Chapel. Joseph appears suddenly and begins to preach. During his words, noises of riot on the soundtrack and actual voices from all around the space gradually get louder, until he is drowned out.

Joseph: 'The empire of reason will ever be the reign of peace. This, ladies and gentlemen, will be the happy state of things distinctively and repeatedly told in many prophesies, delivered more than two thousand years ago; when the common part of mankind will cause wars to cease to the ends of the earth, when men shall beat their swords into ploughshares, and their spears into pruning hooks; when nations shall no more rise against nation, and when they shall learn war no more.' (Isaiah ii 4; Micah iv 3). This is a state of things which good sense, and the prevailing spirit of commerce, aided by Christianity and true philosophy, cannot fail to effect in time. But it can never take place in the wretched manner in which they now are. For peace can n...never be established but u...upon the e...extinction of the c...causes of war.

Voices: *Actors bang on three large planks, and/or bang the planks on the ground, as well as talking and shouting, with recorded sound of the protest in the background.*

Church and King! Church and King! (*Constant, beginning as a whisper*)

Have you seen him? Doctor Phlogiston? We want Gunpowder Joe.

I smell a bad smell ... I smell Joseph Fungus

Where is he? The unchristian ... the Lucifer ... the Cromwell ... the antichrist ... the flaming incendiary ... the Frenchie ... the revolutionary!

Have you heard his arrogance? His egotism?

His unpardonable indecencies?

Where are you Priestley, you political priest? ... you High Priest of Satan ... you slime of the Stygian lake.

Voices: We're coming for you Joe. Lynch him. String him up.

Projection of pulpit fades. Joseph shakes and hides. The voices fade. Mary enters, as if looking for someone in a house.

Mary: Joseph? Joseph?

Joseph: (*Hidden*) M ... Mary?

Mary: Will you come and drink port with me?

Joseph: I was laying more wood on the fires in the upper rooms ... I sat down ... must have fallen asleep ... I thought I heard voices far off ... coming from the direction of the city ... strange cries and huzzas.

Mary: Join me for a glass of port; it will settle your mind.

Joseph: Very well ...

She pours a port for him.

Mary: I'm so glad you stayed here and did not attend the commemoration of the revolution.

Joseph: (*Raising his glass*) To the blessings of equal government in a truly enlightened nation! I should have been there but for stirrings among the ignorant drolls of the king and his bishops. And if I had attended, they would have told the world Sir Lucifer sat at the head of the table.

The voices begin again, faintly. A loud knocking on one of the planks, echoed by knocking on the other two.

Martha: Mary ... Joseph ... are you in there? Come quick.

Mary: Martha Russell. Good evening to you Martha. Will you –

Martha: The mob are heading this way. They are at the work of destruction. They have burnt down the doors at the New Meeting House and made a bonfire of the pews. The Old Meeting House is torn to the ground and now the intoxicated mob cries for your blood, Doctor Priestley. They have set off from Birmingham to find you here.

Joseph: We will put all the fires out, tell the neighbours ...

Martha: You must haste. Mr. Ryland has a chaise waiting.

Joseph: I must pack some clothes, my instruments –

Martha: You must come now-

(*To audience*) We took the chaise but Doctor Priestley did not let us get far. He said the mob was on foot and would not know how to find him. We went no further than a mile before he and Mary walked back to listen to the sacking of their house. It was a clear night, a remarkable night, with a bright moon. We could hear voices from across the fields quite distinctly.

The planks are used to make smashing and breaking noises.

Voices: We have no fire. Two guineas for a candle. They say he has electrical machines. Let's see if we can make them spark. Burn it to the ground! Burn, Burn!

Joseph: Doctor Gunpowder they call me yet cannot set fire to my house!

Mary: I suspect it is the cellar that is getting most attention. They are in a state of bestial drunkenness. Joseph ... let us go further away ... please! I cannot bear the sound of our windows and furniture breaking.

Projection of the engraving of Priestley's house being burnt, with flickering flames-either animated on the projection, or using the stage lighting.

Joseph: Flames at the back of the house. The fire begins!

Mary: They are strewing your books along the highroad. Your instruments ... the laboratory ... the rain gauge ... our magic lantern ... our home! (*She cries*)

Joseph: Things of this world Mary.

Martha: (*To audience*) He just stood there, comforting his wife, while the pages of his beloved books were flying across the fields. He just stood there, listening to the dreadful and hideous shouts.

Martha: The battering of the walls ... and he watched the flames, as if he was peering into a crucible ... so still and attentive ... I thought: his life's going up in flames and yet he makes no complaint, no sigh. He seemed like a sort of child, an innocent staring in wonder.

Joseph: (*Almost to himself*) "We are as it were, laying gunpowder grain by grain, under the old building of error and superstition, which one single spark may hereafter inflame, so as to produce an instantaneous explosion."

	I w ... w ... wish I had not said those words. They were ill advised. I did not wish any violence.
Mary:	Come away Joseph ...
Martha:	More hundreds come to the mob from the town ... into the chaise quick. We must get more distant.

The voices rise to a crescendo before suddenly stopping.

Joseph:	I thought ... it will pass over. I planned to return to the charred remains of the Meeting House and preach a sermon on the Sunday in the open air. "Father forgive them for they know not what they do." But they pursued me. Their voices still echoed in my head when we arrived in London. I was an object of troublesome attention wherever I went, wherever I preached. Outside my own circle, I felt only unkindness and hostility. When war broke out with France, I feared I should be sent in irons to Lancaster gaol, so we said farewell to our friends, and embarked on the Sansom at Gravesend, bound for New York.

Scene 2. On board the Sansom .

The planks are used to suggest the motion of the ship and the waves. A projection of waves. Mary is lying down. Joseph enters with a candle.

Joseph:	Mary?
Mary:	I am so sick ...
Joseph:	This day I saw mountains of ice.
Song:	The ship it rocks by night The ship it sways by day but Joseph loves the sea, experimentally!
Joseph:	I have been testing the heat of the water at various depths and made observations towards experiments.
Mary:	I have experienced sickness for three weeks ... surely a record.
Song:	The ship it rocks by night And the ship it sways by day but Joseph loves the sea and writes on infidelity
Joseph:	I have finished my sermon: "Observations on the Cause of the

present Prevalence of Infidelity" I have completed my reading of Hartley's second volume and today I performed divine service for the steerage passengers.

Mary: My stomach is ... shipwrecked!

Song: The ship it rocks by night
The ship it sways by day
but Joseph loves the sea
for its divinity

Joseph: Tonight Mary, I shall read you the tale of Jonah and the Whale. Jonah, as you will remember was thrown overboard and sank –

Mary: No, no, go away. Tell the captain we must stop.

Joseph: The Captain is disorderly, my dear. I fear he is much given to liquor. But today he came jovially to tell me the topsail has blown away in last night's gale. He rejoiced that we have not yet been boarded by privateers or shot across our bows by a vessel of the French.

Mary groans.

Song: The ship it rocks by night
The ship it sways by day
but Joseph loves the sea
and Mary loves the sea
unexpectedly

Mary: After six weeks, I think my sickness has finally departed. Let us walk on the deck, Joseph, and enjoy the stimulation of the ocean air.

Joseph: I fear I am too busy with my writings, my dear. Longhand is laborious in these conditions.

Song: The ship it rocks by night
The ship it sways by day
And they both love the sea
but separately!

Joseph: Mary, today I have seen great spouts of water –

Mary: Do not mention spouts!
Oh no ... not again!

Joseph: Land ahoy my dear ...

Scene 3. Northumberland, Pennsylvania, America

Projection of forest, Joseph and Mary are balanced precariously on raised planks.

Joseph: The creeks were swelled with rain but we and the baggage were ferried over by canoe. Mary was much discomforted but I thought myself to be Robinson Crusoe coming upon a deserted island from the rough billows. In the end, after enduring bed bugs and two nights in the back of a wagon, we came to Northumberland in Pennsylvania, at the point where the northern and western branches of the Susquehanna meet. It was almost the end of the known world. Our new place of abode was but a settlement in a forest clearing.

Mary gracefully descends from the planks, helped by Joseph.

Mary: I am so happy and thankful to meet with so sweet a situation and so peaceful a retreat.

Joseph: Here, I trust, we shall have no apprehension of metaphorical gunpowder being mistaken for real. I intend to travel to Philadelphia to preach for two months of the year. And when my books and instruments arrive, I shall prove beyond any doubt my theory of phlogiston and complete my Church History.

Mary: (*Looking around*) This is a fine site on which to build our wooden house. I do not think there can be a more delightful situation than this in the whole America.

Joseph: (*Picking up the planks*) We commonly kiln-dry ten thousand feet of boards at a time. The firewood must be such as is not apt to flame, lest the boards should take fire. Our sons are taking to farming like ducks to water especially our youngest – Harry. I too sometimes take my mattock and work alongside them but I must make haste to Philadelphia. I have great hopes of founding a Unitarian Society in that city.

He puts the planks down and leaves.

Mary: Joseph … Joseph! … Joseph!!!

She picks up a plank.

(*To audience*) There was much work to be done. Sufficient planks had been prepared to complete the main rooms of the house but Joseph remained in the city for nearly five weeks and when he returned, he wished only to converse about the success of his favourite pursuit.

Joseph comes back in, talking excitedly.

Joseph: Mary, Mary, I preached on many occasions to a thronging multitude and, I think, gave universal satisfaction. The Vice-President, Mr. John Adams, was most punctual in his attendance. And to think only five years ago a Unitarian who, like ourselves, denied the Divinity of Christ, was forced to flee the city, I have great hopes of my next visit.

Mary groans, a knocking on a plank.

Mary: Martha... old friend... how very delightful to see you after all this time.

Joseph: Ah, a fellow sufferer!

Mary: (*laughs*) Joseph has given up his wig and his teeth are falling out!

Martha: Fortune has not favoured us my dear Mary. Our ship was captured by the French. After five months imprisoned in ships at Brest, I am afraid we have become most critical of the French Revolution.

Joseph: Hrrrumph!

Mary: (*laughs*) Remember that Joseph was made an honorary citizen of that country.

Joseph: We shall be most pleased to have your family in close proximity to us in Northumberland.

Martha: I think we have settled on Connecticut.

Joseph: Huh ...

Martha: The younger members of the family are quite decided.

Joseph: We have much wood for house building here-

Mary: Joseph!

Martha walks away and then stops, leaving Joseph and Mary looking at her.

Martha: (*to audience*) It was soon after my visit that Harry, Mary's youngest son, and Joseph's favourite, died, supposedly from overtaxing his strength.

Martha leaves

Joseph: (*Holding a plank*) Then my beloved Mary died, shortly before our house became habitable, broken by the loss of our son.

Joseph: (*Looking at plank*) On her gravestone, I had engraved: "God shall wipe all tears from their eyes and there shall be no more death, neither shall there be any more pain: for the former things are passed away."

He puts the plank down

Never did my trust in Divine Goodness weaken, yet walking in the woods, among the aspen leaves of autumn, sometimes made me wish that I would lie beneath them, buried with my beloved Mary, though I hoped to visit England once more.

He lights a candle
Echoes of the voices: "Doctor Phlogiston, Gunpowder Joe" etc, which continue beneath the following lines:

I rise first and light the fire each morning.

I still conduct experiments.

I have found favour with the government here.

The voices stop.

I feel so... old. Grandchildren! (*More feebly*) Grandchildren...

Three actors (playing the three young women) come to kneel before him.

I am going to sleep as well as you – a good long sound sleep. We shall meet again.

When I exposed hydrogen and air to an electric spark inside a vessel, water was deposited on the walls. If we can make water, who knows what is possible.

One of the actors blows the candle out. The three phrases below overlap and repeat three times, as Joseph leaves very slowly.

Grandchild 1: Death came so peacefully.

Grandchild 2:	So peacefully.
Grandchild 3:	A good long sound sleep.
Grandchild 1:	We shall meet again.

Part 2: Return for one night only.

As Joseph disappears, a projection of his portrait comes up, which gradually disintegrates as the noise of the city rises on the soundtrack:voices, cars, police sirens, the screeching of brakes, breaking glass etc. The noise suddenly stops.

Scene 4. In Mill Hill Chapel, Leeds. Spring 2004.

Projection of the interior of Mill Hill Chapel, showing the altar (not necessary in the chapel itself). Fina enters from the back of the space, carrying a decorative candle. She looks at it – and the other two use it as a focus as well.

Fina:	I woke up this morning and I thought it might just all have been a dream but ... I looked at your bed ... and I remembered ... I remembered the sound of your scream ... the screeching of the car ... the sound of the ambulance ... tears running down my cheeks ... holding your hand. (*Angry*) He never stopped to look!
Jenny:	She's gone, isn't she. She's never coming back.
Lucy:	I just want to tell you that I loved you. We all loved you. All them good times, they'll never be forgotten.
Fina.	I don't know why I'm here, Louby I don't know where you've gone to now but I was walking through town ... and this place drew me ... I mean I must have seen it loads of times, but I never really noticed it ... not till today, I bought a candle ... for you. You used to light candles in your bath.
Jenny:	That car ... it were going so fast. He just didn't see her.
Fina:	He didn't care. (*Sudden loud sob*) Why didn't he stop?
Lucy:	We'll never forget her, will we?
Jenny:	Never ... Shall we light it?

She lights the candle and they all look at it. Joseph walks slowly to his chair.

Lucy:	Where shall we put it?
Fina:	Lets put it in the best place, where everyone'll see it.

Jenny and Lucy look blank.

Right on the altar.

She sees Joseph and drops the candle. Stifled scream – they freeze. He feels the arms of the chair.

I shall not hurt you. My name …

My name is Joseph.

This chair … this is my chair, in Leeds, at Mill Hill Chapel … but this does not look like my-

Lucy: He's just one of them old scruffs that hangs around – been sleeping it off on one of the pews. We just didn't see him.

Joseph: My name is Joseph Priestley, one time minister of this chapel, originator of the Phlogiston theory, author of –

Fina: Joseph Priestley! No way: he died two hundred years ago.

Lucy: What?

Joseph: My name *is* Joseph Priestley.

Fina: (*Whispers*) Joseph Priestley discovered oxygen … well, sort of.

Lucy: (*Boldly*) Where have you come from then, Joseph?

Joseph: Northumberland, Pennsylvania … no, no … before … long before. I hear them … I hear them in my head.

Jenny: What's he doing now? It's like he's listening … but not to us. He's really weird. I'm going to get out of here.

Fina: Sssh!

Lucy: (*Whispers*) Let's go along with it a bit. He seems harmless enough.

Mother: Joseph, let us say together, "Our father which art …"*

Joseph: Mother? I can smell cloth. My father was a maker and dresser of woollen cloth at Birstall Fieldhead, six miles from Leeds.

Timothy: Joseph.

Joseph: Ah, Timothy my brother.

Timothy: Do you remember the spiders?

Joseph: Ah, the spiders. I put them into bottles and found out how long they could live without fresh air.

Aunt Sarah: Joseph, It's time for the meeting of young men, for conversation and prayer.

Joseph: Aunt Sarah. She was truly a parent to me after my mother died. A strict Calvinist, and in all respects a perfect human character.

Timothy: Hey Joe, let's go rabbiting.

Joseph: No.

* The voices of Joseph's mother, Timothy and Aunt Sarah are pre-recorded.

Timothy: (*chants*) Joseph is boring, Joseph is boring.

Joseph: I have a sermon to copy and Mr. Kirkby has set me some Hebrew verbs to learn.

Timothy: Boring, boring!

Aunt Sarah: Joseph you are ill.

Joseph: I cannot breathe, Aunt … I think I shall not be long lived. I feel occasionally such distress of mind … I am like the man in the iron cage in the 'Pilgrim's Progress'. But these conflicts of mind have their uses. They lead me to think of God and a future state.

Fina: I don't think I ever think of God.

Pause

Joseph: They are gone and now I have returned for some great purpose. I did rejoice in the firm belief and expectation of my body rising again incorruptible at the last day.

Fina: If you are Joseph, you were a scientist. You can't believe in all that God stuff.

Joseph: The true scientist is a religious man –

Lucy: Or woman.

Joseph: Devoted to his task out of a supreme veneration for the God of nature and from a love of his fellow creatures.

Jenny: Well, I'd like to believe we'll see Louby … see Louise again, not all broken and mangled.

Lucy: Our friend died. She was just twenty. Your science can't do anything for her … or your God.

Joseph: Light your candle. I have conducted many experiments with candles … and mice.

Jenny: Not mice. I hate mice. They freak me out.

Fina picks up the candle and lights it. Projection of Joseph's scientific equipment – 3 engravings are shown during the following discussion.

Joseph: I kept mice on my kitchen range, kept them warm in my Yorkshire kitchen.

Jenny: That's cruel.

Joseph: Please (*He takes the candle*)

I employed various sources of heat, from the range, from candles, and once I obtained a convex lens to see if I might harvest the heat of the sun, focusing it on substances in the hope that they would generate new gases. When I did this with mercurius calcinatus –

Lucy: God, what's he on! Translation please, Fina: you're the scientist.

Fina: Sssh!

Joseph: I found that an air was released and what most surprised me was that when I put the candle in it, it burnt with a remarkably vigorous flame.

Fina: Because it was burning in oxygen.

Joseph: No, no … dephlogisticated air.

Fina: (*mutters*) Same difference.

Joseph: (*He holds up the candle*)

But imagine if I had not had a lighted candle nearby … the whole train of experiments might have been stopped. And I can tell that in this dephlogisticated air my mouse lived a full hour but had it been common air, it would only have lived for a quarter of an hour.

Jenny: So the mouse died.

Joseph: No, no my dear. It was exceedingly chilled, but upon being held close to the fire, presently revived.

Fina: Yeah, 'cos you put it in oxygen. You discovered oxygen.

Joseph: No … I discovered dephlogisticated air.

Jenny: Might as well be supercalafragilisticexpealidocious air. I just don't get it … an' I don't get him

Lucy: Me neither.

Fina: Planks over there.

Lucy: What?

Fina: (*Sighs*) Give me the planks, (*Takes one plank*) This is mercury. (*Takes another plank*) This is oxide … put the two together and you've got mercuric oxide.

Joseph: Mercurius calcinatus.

Fina: That's what I said. Last plank. (*Takes plank*) What he thought was that when mercuric oxide burnt, it took up this stuff from the air called phlogiston and turned into metal-into mercury.

Lucy: Three planks stuck together. Great!

Fina: It's hard to get your head round. If you take the phlogiston out of the air, you get dephlogisticated air, which is what the mouse was in …

Joseph: An excellent elucidation ...

Fina: But I haven't got another plank for dephlogisticated air.

The last projection of equipment fades out.

Jenny: I still don't get why the mouse didn't die.

Fina: Because it was in pure oxygen.

Joseph: When the candle burns it releases phlogiston, which makes the air less springy.

Fina: Or to put it another way, oxygen is used by the combustion or respiration, the mouse was breathing in oxygen and –

Joseph: Oxygen, oxygen! I am a phlogistonist.

Jenny: Too much. Too much. I want to put the candle on the altar, for Louby, and get out of here.

Fina: Me too ... but it's like ... it's like because of him that they give you gas and air when you have a baby ... and they gave it to Louby in the back of the ambulance.

Jenny: 'Cept Louby died.

She cries

I miss her so much.

Fina and Lucy comfort her. Joseph blows out the candle.

Joseph: For now, your friend is gone and sleeps soundly, I must do what I was brought back to do. I shall preach. I shall preach to this generation of the glory of God and his creation.

Projection of the pulpit at Mill Hill as he makes his way to the pulpit.

Fina: (*Groans*) Religion, it's all rubbish.

Jenny: Sssh

Joseph: 'And God saw everything that he had made, and, behold, it was very good.' Genesis 1, verse 31.

The connection that all persons, and all things necessarily have, as part of an immense, glorious and happy system (and of which we ourselves are part, however small and inconsiderable) with the great author of this system, makes us regard every person, and every thing, in a friendly and pleasing light. The whole is but one family. We have all one God and Father ...

Voices start, at first tentative, then louder, drowning out the sermon, using phrases like: Bible Basher; Jesus Freak; Geek; God Squad; The Darkness are God; Ozzie's God; Jack's Jesus. Joseph's sermon disintegrates until the noise suddenly stops.

whose affection for us is intense, impartial and everlasting. He despises nothing that he has made not one of you here today and by ways unknown to us and often by methods most unpromising, he provides for our greatest good. We are all training up in the same school of moral discipline, and are likewise joint heirs of eternal life, revealed to us in the gospels.

Silence. Joseph holds his head in his hands.

Jenny: He's hearing voices again; It gives him some kind of pain.

Joseph: (*Shouts*) Where is my liberal, friendly and harmonious congregation? Where are my happy years here? Where have they gone?

Timothy: Joseph, the people don't want you. They're not listening*

Joseph: Timothy?

Aunt Sarah: Did we make you a scholar for this Joseph?

Mother: I lost so many children Joseph. Go now, or they will take you too.

Joseph: Mother?

* The three voices of Joseph's relatives are pre-recorded again

The city/clubbing noises begin again.

Lucy: Go home Joseph, wherever that is.

Joseph: England once more. I was to come to England once more.

Fina: It's not your England, Joseph. But people remember you: you're sort of famous.

Joseph: But do they hear His word?

Lucy sighs.

Jenny: We came here to remember our friend ... Louise. We ... we don't really want you here now.

Lucy: Will you go, if ... if we say a prayer for you?

Joseph: The first prayer. My mother taught me. 'Our father which –

Lucy: Yeah, that's the only one I know ... and I don't know if I'll remember all of that. Fina? Jenny?

Fina: OK.

Lucy: Never thought I'd be doing this.

Joseph slowly and quietly walks out during the prayer. City noises continue quietly in the background.

Our father, who art in heaven –

Fina: And Louise if you're up there too.

Lucy: Hallowed be thy name, thy kingdom come.

Fina: I wish you could come back Louise.

Jenny: We miss you so much already.

Lucy: Thy will be done, on earth as it is in heaven.

Jenny: What's it like up there?

Fina: Joseph's going. Come on, Lucy.

Lucy: Ur … give us this day our daily bread.

Jenny: And chocolate spread!

Fina: Sssh!

Lucy: And forgive us our trespasses, as we forgive those who trespass against us.

Could you forgive him? The man that did this to you? I don't know if I could.

Lead us not into temptation, but deliver us from evil –

Fina: He's gone. Joseph's gone.

Lucy: For thine is the kingdom –

All: The power and the glory, for ever and ever. Amen.

The End

Woodkirk School Cast and Crew Pictures

Lloyd Bradley and Rachel Harrison as Dr. and Mrs. Priestley *(below)*. Shot in Mill Hill Chapel after the dress rehearsal.

(above) Cast, Crew and Woodkirk High Specialist Science School Staff at Mill Hill Chapel, Leeds, behind the 'Priestley' Chair.

From left to right: Virginia Stokeley (Producer), Rev. Paul Travis (Minister at Mill Hill Chapel), Adam Strickson (Playwright and Producer), Fay Williamson (Fina), Rebekah Cohen (Martha Russell/Lucy), Lloyd Bradley (Joseph Priestley), Rachel Harrison (Mary Priestley), Andrew Hutchinson (Narrator), Kelly Mallinson (Jenny), Joe Wretham (Composer/Sound), Emma Lawton (Director), Alex Baker (Composer/Sound), Marie Webster (Assistant Producer), Bill Bailey (School Headteacher).

[Not present: Alex Walmsley (Lighting/ Technical Assistant), Lynette Haddock, Alex Merson, Sonia Mills, Chris Shaw, Peter Jackson, Alison Balmforth, Samantha Waterfall (Researchers), Fiona Connelly (Costumes), Jonathan Birch (Technical Director), Chris Boylan (Visual Production).]

Rebekah Cohen as Martha Russell, a friend of the Priestleys, (in the presence of Joseph and Mary Priestley) describes the destruction of their Birmingham home.

Performance at Mill Hill Chapel. The screen was used to project scenes from Priestley's life and diagrams of his apparatus during the action. The atmosphere of the chapel made some of the speeches by Lloyd Bradley (Joseph Priestley), designed to convey some of Priestley's religious convictions to a modern audience, one of the most powerful performances he gave.

SECTION 6

July 4th Priestley Lecture

PRIESTLEY TO AMERICA AND THE FOUNDING FATHERS

Priestley Toulmin
Great-great-great-great grandson of Joseph Priestley

Priestley Toulmin

Prologue

I am greatly honoured – and somewhat abashed – to have been asked to inaugurate what is certain to be a distinguished and prestigious lecture series. Though I may claim some standing as a geologist, a geochemist, and perhaps an amateur genealogist, I am certainly no historian. I suppose I am here by genetic happenstance. But there you are, and here am I, and we must muddle through together.

Introduction

Joseph Priestley spent the last decade of his life in the United States of America, the upstart country whose revolutionary origin he had strongly supported from his native England.

That support was but one – and by no means the most important – reason for the persecution that drove him somewhat reluctantly from the country he never ceased to love. Priestley had for years been an enthusiastic supporter of the democratic (or "republican", as they were then also called) principles inspiring – as he thought – both the French and American Revolutions, and his continuing approbation of the French Revolution, even as he deplored its later excesses, was a major factor in his unpopularity in many circles both here and later in America. His unorthodox religious views and fervent opposition to the very concept of an established church brought down the wrath of ecclesiastical authority on his head. His outspoken views on authoritarian government incurred the severe displeasure of the political establishment. Both factions energetically, if sometimes surreptitiously, encouraged measures designed to diminish his comfort, happiness, and influence. Antagonism was reflected in popular media (Fig. 1) and extended to his family as well (Fig. 2).

Fig. 1. *(left)* "Doctor Phlogiston" cartoon
ARTIST UNKNOWN

Fig. 2. *(right)* William Priestley being presented to the French Assembly
JAMES SAYER

So Joseph Priestley finally decided to emigrate to America in the fall of 1793. Although the thought of removing to America had occurred to him as early as the 1770s, the Birmingham riots of 1791, in which his home and belongings were destroyed (Fig. 3) and from which he and his family barely escaped unharmed, and the increasing hostility to him and his family over the ensuing years, when he had established himself in London, finally forced him to a reluctant decision to leave his native land.

He had waffled in indecision whether he should seek refuge in France or in America, but perhaps because of the turn of events in France, more likely because of his sons' decisions to move across the Atlantic, he settled upon "retirement" in America, where he expected – and to a considerable degree found – a more hospitable environment. He could (with his wife's major assistance) build and reside in a comfortable residence (Fig. 4) unthreatened at least by physical violence, in a land that conferred many honours, and continues to do so, even to recent days, as shown, for instance, by this postage stamp, issued a few years ago (Fig. 5). They could have picked a more flattering image, I

think. Although this was probably not a deliberate slur, the engraver certainly exercised the full measure of artistic license if as I understand the figure on the stamp was derived from the Peale portrait (Fig. 6). In any event, I prefer the Gilbert Stuart rendition (as did Priestley) (Fig. 7).

Fig. 3. Destruction of Priestley's home in Birmingham, 1791

Life in America

Let us now turn to the story of the Priestleys' emigration and subsequent career in America. The skeletal timeline in Fig. 8 may provide some chronological mileposts along the way. It also inspires a small digression to emphasize the point that though many think of Priestley primarily in terms of his chemical pursuits, secondarily perhaps as a political and theological radical, he also made major contributions to other fields, including education. He used timelines, too, in teaching history, biography, and other subjects (Fig. 9). But back to our story.

Fig. 4. Priestley's house, Northumberland, Pennsylvania

Fig. 5. (*above*)
Commemorative
U. S. postage stamp.

Fig. 6. (*right*)
Peale portrait of Joseph
Priestley

Fig. 7. Stuart portrait of
Joseph Priestley
COPY BY A. ROSENTHAL

1789	WASHINGTON President, ADAMS Vice-President
1790	FRANKLIN dies
1791	July 14 Birmingham riots.
	Priestleys to London
1792	WASHINGTON, ADAMS re-elected
1793	April–AugustWilliam Priestley leaves France for America
	August Joseph Jr. & Harry leave for America
1794	April–JunePriestleys to New York, then Philadelphia
July	Priestleys to Northumberland
1795	December Harry Priestley dies
1796	September Mary Priestley dies
1797	ADAMS President, JEFFERSON Vice-President
1798	June–July Alien & Sedition Acts
1801	JEFFERSON President, BURR Vice-President
1804	February 6 Joseph Priestley dies

Fig. 8. Significant dates in Priestley's American life

In April of 1794 Dr. Priestley and his wife sailed from Gravesend in the good ship *Sansom*, arriving in New York some eight weeks later on June 4. Their sons had preceded them. Both Dr. Priestley and his second son, William, had been accorded French citizenship, and the father was appointed to the Conventional Assembly, though he declined the honour. William had been in Paris in 1789, and may have observed the storming of the Bastille. He was in Birmingham at the time of the riots in '91, where his conspicuous and active role in the defence of his parents' house had rendered his continued residence

Fig. 9. Priestley's "timeline of history"

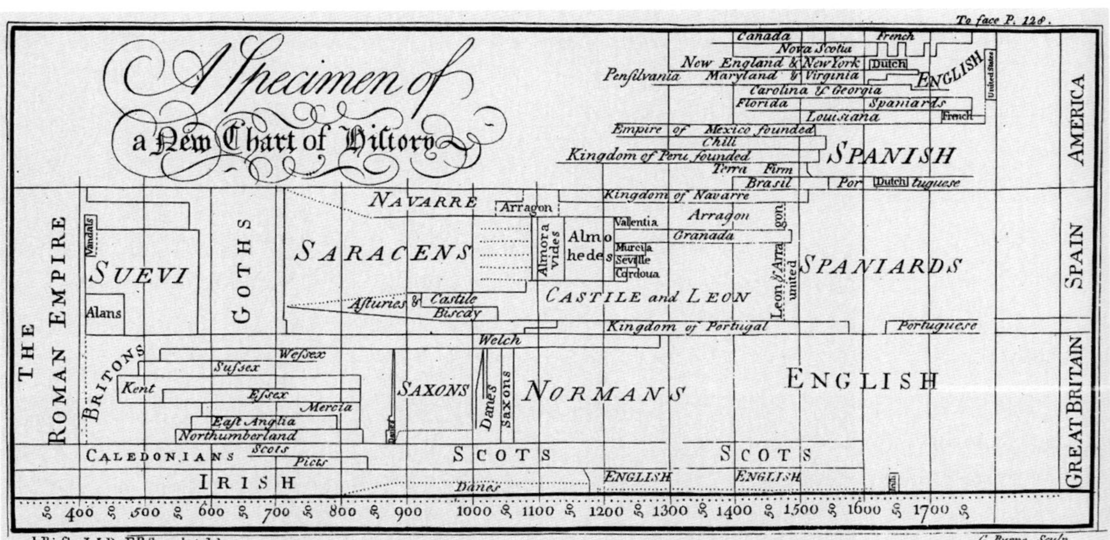

A New Chart of History
by Joseph Priestley

Joseph Priestley House
Northumberland, Pennsylvania
Penna. Historical & Museum Commission

there unwise. In any event, by the summer of '92 he was settled in a banking house in Nantes, and as mentioned had accepted French citizenship. Within less than a year, however, conditions in France had deteriorated to the extent that he decided to remove to America. He arrived in Philadelphia in the summer of 1793, at which time his brothers were *en route* as well, having made their decision in March.

Priestley's eldest son, Joseph, had been employed in a Birmingham mercantile firm from the spring of 1791, but I do not know what role he may have played during the riots in July. About a year and a half later, however, in the winter of 1792–3, he was compelled to abandon his partnership there, and the following summer he and his younger brother Harry sailed for America. Joseph, junior, was – at least at this time – a strong supporter of the French Revolution and other radical liberal causes, and was inclined more to seek refuge in France than in America, but was convinced by family and friends that America was a more suitable retreat. He and Harry sailed for America in August 1793, in the company of Thomas Cooper, a close family friend, controversial activist, barrister and chemist, of whom more will be said later.

Harry, not quite 16 and a student at the Dissenting Academy at Hackney, where his father was teaching at the time, had, in his father's words, decided upon "an active employment, rather than a profession" and was to be looked after by family friends in America, the Vaughans, and prepared for an agricultural or commercial career. The two brothers and Cooper sailed (from Liverpool?) in late summer of '93. They were at first undecided as to their final destination in America, planning to visit Kentucky with Harry Toulmin, a family friend, whom they planned to meet in Philadelphia, but also considering Virginia (at the urging of both Jefferson and Madison) and Boston. In the end, however, Joseph junior and Cooper decided upon an ambitious scheme to acquire land in Pennsylvania and establish an émigré community there.

Dr. and Mrs. Priestley were met in New York by their son Joseph and his wife, and were accorded an extravagantly flattering welcome by the powers that were in the city and region. Governor Clinton (George, no relation I believe to the present-day political figure), and many dignitaries of the educational and mercantile establishments joined in welcoming the great man to the new nation. The contrast with the treatment he had become accustomed to in his native land was not lost on Priestley, who noted that "In England, I was an object of the greatest aversion to every person connected with Government; whereas here, they are those who show me the most respect". The enthusiasm of his welcome was presumably occasioned by his fame as a scholar and intellectual and his reputation as a supporter of the cause of liberty and the American movement to independence. To have so noteworthy a person choose to reside in the new country was a cause of satisfaction to many. The local ecclesiastical community, however, were somewhat less thrilled. The preachers of the city were for the most part, he said, civil to him, but none invited him to their pulpits. This cautious – at best – reaction in certain quarters should not have been a great surprise to Priestley, for he had been receiving for twenty years or so gentle warnings from knowledgeable correspondents in America that the intellectual climate there was not wholly free of bigotry and illiberality. For the time being, at least, however, Priestley seems to have tried carefully to avoid overt political involvement in the affairs of his new country, but he could not avoid for long the association of his philosophical ideals with direct political consequences, and the attendant controversies.

In any event, after about a fortnight in New York, the Priestleys, including Joseph junior and his wife, moved on to Philadelphia, at that time the site of the Federal government, where he again was received with much flattering ceremony, though apparently the religious community was no more welcoming than it had been in New York. Again he was not invited to preach in any of the local churches, but he received an enthusiastic 'Address' of welcome from the American Philosophical Society, an honour that meant much to him. In response, he expressed his view that the republican form of government, as represented in his view by the United States and revolutionary France, would be " ... far more favourable to science and the arts than any monarchical government has ever been". Despite his oft-repeated declaration (at this time) that he had no wish to inject himself into the highly contentious American political maelstrom, we see that Priestley was as usual unable to restrain himself from the forthright expression of views that might not sit well with some parties in the fray.

The Priestleys stayed in Philadelphia for a month. During their time there, Priestley became well acquainted with David Rittenhouse, President of the Philosophical Society, and Benjamin Rush, a prominent physician and statesman, and signer of the Declaration of Independence. He socialized extensively with John Vaughan and his Federalist circle of friends, but clearly preferred the anti-Federalist, or republican, political philosophy.

From Philadelphia the Priestleys moved on to the small frontier community of Northumberland, at the picturesque forks of the Susquehanna River, in order to be closer to the site of the emigré community planned by their son Joseph and Thomas Cooper. They – or at least Dr. Priestley – at first seemed to have planned to make Northumberland more of a retreat than a principal place of residence, which he thought of as Philadelphia, but very shortly he was writing to correspondents that he planned only extended visits to Philadelphia, generally to maintain intellectual and social contacts while Congress was in session. He was at this time sanguine about the prospects of establishing an Academy, or college, in Northumberland, for which he was laying ambitious plans. His major lament was the isolation of the place and the poor communication with Philadelphia and the outside world, *i.e.*, England and the European Continent. The post was infrequent and expensive, and haulage beset with many difficulties, making the establishment and furnishing of the house and laboratory he intended to build difficult. He appealed to John Adams, then Vice-President, with whom he was still on friendly terms, for help in establishing regular postal service to the town, but it is not clear to what degree the service improved as a result. Nonetheless, he remained optimistic about the prospects both of his academy and – briefly – of the emigrant colony that would nurture it.

The colony plan collapsed fairly abruptly in the fall of 1794, however, victim of a combination of factors. As Schofield and others have pointed out, this was probably just as well, as few of the intellectually rebellious sophisticates who might have come possessed the temperament or skills to prosper in the wilds of Pennsylvania's forested mountains. This setback, combined with the aforementioned difficulties of isolation, must have made the offer of the Chair of Chemistry at Philadelphia mightily tempting, and in fact Priestley wrote rather poignantly, "Had this proposal been made to me before the removal of my library and apparatus hither, the case would have been different; but this being now done, at a great risk and expence, I am, at all events, fixed for the remainder of my life". [Priestley to Lindsey, 14 September 1794, quoted by Graham, p. 64]. Many of his contemporary friends, and subsequent writers on his life, have speculated on the reasons for his choosing to remain in an environment seemingly so incompatible with his interests and needs. In addition to the factors already mentioned, simple economics and domestic tranquillity perhaps played a part – living expenses were much less than in a large city, and Mrs. Priestley had conceived a strong distaste for urban life after the experience of the Birmingham riots.

Gradually adapting to this life of rustic isolation, Priestley – or more accurately, his wife – began planning the construction of his new house and laboratory. Progress was agonizingly slow; labour was scarce and poorly skilled, materials expensive and difficult to come by. Well-seasoned lumber was unobtainable, and wood had to be kiln-dried on the site. Most of 1795 seems to have been devoted to planning and the writing of philosophical tracts, though some experimentation was carried out in a separate room in his son Joseph's house, near where the family was living pro tem. Ill health began to afflict the family, however. The youngest son, Harry, had been subject to attacks of fever since the summer of the family's arrival in 1794, and died in December of 1795. His death was deeply felt by both parents. His mother, meanwhile, began showing symptoms of tuberculosis, and did not live to see the house she had spent so much effort in planning and designing.

Priestley began his more or less regular visits to Philadelphia with a three-month stay in February to April 1796. He was by then in a more optimistic mood. The city generally welcomed him warmly, though the generality of preachers continued to refuse him their pulpits. He nonetheless arranged to give a series of lectures, the "Discourses on the Evidences of Revealed Religion", in the Universalist church. These lectures were extremely well attended, to the point of filling the hall so that people were obliged to stand in the aisles and doorways. The audiences included members of Congress and even the Vice-President, John Adams, to whom Priestley in deference dedicated the published collection of addresses. On this visit he also became somewhat acquainted with George Washington. An indication of his mood at this time is provided by his admiring assessment of the country and its political process: he describes the debates in Congress, many of which he attended, as eloquent, well-reasoned, and without rancour, with harmonious social relations among the members upon resolution of contentious issues, all of which he contrasted favourably with current practice in Parliament. In all honesty, even tonight on the Glorious Fourth, I find it a little hard to accept this assessment at face value. By all accounts, the political environment in America, as in this country, was more than a little heated at this time. The session of Congress that Priestley witnessed voted to ratify a treaty with England that sent republicans into paroxysms of fury by its favouritism for England over France. Thomas Jefferson, shortly to become Adams's Vice-President after losing a close but relatively amicable battle for the presidency, pronounced it " ... an infamous act ...", and John Jay, the treaty's author, was burned in effigy in numerous venues that probably did not qualify as sites of reasoned debate.

The next four years were not particularly happy ones for Priestley. His beloved wife Mary had died in December 1796. His second annual visit to Philadelphia, in the spring of '97, was less successful than the first. He again delivered a series of lectures – the second part of his "Discourses" – but they were not nearly so well attended as the first. The President (Adams) did not attend. Politics was heating up, with an ever more antagonistic party system developing, and Priestley, constantly disavowing any involvement in American politics, just couldn't keep his strong opinions to himself. William Cobbett, better known on this side of the pond as a radical reformer in the cause of the rural poor, but at this stage of his career an equally radical monarchist in America, subjected Priestley (and other liberals) to a never-ending barrage of scurrilous attacks, eventually goading Priestley into an ill-considered public response. Indeed for the next few years he repeatedly expressed himself forcefully in private and in public, and finally published a series of Letters that went well beyond self-defence to roam over the broad field of American politics in a mode severely critical of Adams and his reigning Federalist party. One wonders whether President Adams was not beginning to regret his decision of the previous summer not to deport Priestley under the Aliens Act. Even some Republicans regarded his radical ideas as extreme, but Jefferson, uncomfortable in his role as Vice-President to a President with whose policies he strongly disagreed, wrote Priestley in warm support. Nonetheless, for the remaining years of the Federalist administration, Priestley's situation was one of unease and constant though not imminent peril of prosecution or deportation.

With the election of Jefferson, close as it was, in 1800, Priestley found himself "for the first time in my life ... in any degree of favour with the governor of the country in which I have lived ..." [Priestley to Logan 26 December 1801, quoted by Graham, p. 154]. His final years were devoted largely to the writing and revision of a variety of religious and philosophical works, tasks that had been put off during the more turbulent 1790s when he had been to some degree preoccupied with political themes. He continued, doubtless at a somewhat reduced pace, the chemical experiments he had pursued ever since establishing his laboratory in Northumberland. Robert Schofield, in his forthcoming book, "The Enlightened Joseph Priestley" [subsequently published; see Reference List], points out that although Priestley published more scientific papers in America than in all his years in England, the substance was largely, in his word, anticlimactic, and that though most of these were ultimately vain protests against the "new chemistry" of Lavoisier *et al.*, many of the objections he raised were valid at the time, and were only refuted by later experiments and discoveries. Frustrated no doubt by the inattention his pro-phlogiston arguments received, he nonetheless basked in the general approval with which he was now regarded, finding particular gratification in a testimonial dinner given by the American Philosophical Society in March 1803. His health had begun visibly to decline in 1801, when he had become seriously ill on a visit to Philadelphia, and he died peacefully in Northumberland on 6 February 1804.

"Founding Fathers"

Much has been made of Priestley's relationship to those who are somewhat elastically referred to as "Founding Fathers". He had, of course, known Benjamin Franklin well in London, and corresponded with him from his return to America in 1775 until his death in 1791, though obviously they never met in America. Priestley was acquainted with all three Presidents of the United States during his lifetime, and carried on extensive correspondence with two of them. James Madison, whose Presidency occurred after Priestley's death, had, along with Jefferson, strongly urged the Priestleys to settle in Kentucky, as their friends the Toulmins did, and the two may have met personally as well.

Franklin

Benjamin Franklin represented various colonial interests in London from 1757 to 1775, when he returned to America, never to return to England. His association with Priestley began in 1765 when the latter was visiting London from Warrington. By then, Franklin, 27 years older than Priestley, was as prominent a scientist as he was a diplomat, his electrical researches having won him the Royal Society's Copley Medal even before he arrived in England. The older man became the younger's mentor, and encouraged him to publish a compilation of the history of electrical research. When within a year or so Priestley did not only that, but repeated many of the earlier experiments and performed significant new ones, Franklin and others nominated him for membership in the Royal Society, and Franklin went further to propose him for the Copley Medal, which was however unsuccessful [Priestley later (1773) was awarded the medal for his work in pneumatic chemistry]. They saw each other frequently at meetings of the Club of Honest Whigs, a group of congenial liberal-minded intellectuals of various fields that met regularly at a coffee house in London. Priestley and Franklin carried on a voluminous correspondence, ranging over science, politics, and philosophy, both before Franklin's return to America and afterwards, until his death in 1790.

Washington

Priestley first met George Washington not long after his arrival in America. On his initial stay in Philadelphia, then the capital of the fledgling nation, the President received Dr. Priestley and Thomas McKean, the Chief Justice of Pennsylvania at the time, in what was apparently a somewhat perfunctory, ceremonial audience. They met again, at greater length and more intimately, a couple of years later, on one of Priestley's regular visits to Philadelphia from Northumberland. On this occasion Priestley was accompanied by William Russell, a Birmingham friend now settled in America but who had organized the Bastille Day dinner in Birmingham that provided the flash point for the Birmingham Riots of 1791. The visit (or visits – there may have been more than one) was probably social only, though Washington was apparently familiar with Priestley's theological writings, which they discussed. I know of no correspondence between them, at least of a substantive nature. Priestley apparently found Washington a bit over-diplomatic for his taste, for he later wrote that he preferred President Adams's more open – he used the term 'undisguised' – style to that of his predecessor.

Adams

Priestley's relationship with John Adams was more complex. They first met in London in 1786, when Adams was serving as Ambassador to England. Priestley was then in Birmingham, but making regular visits to London to renew acquaintances and keep up with the intellectual world there. After Adams's return to America, they apparently kept up some sort of correspondence, for Adams, by then Vice-President of the United States, wrote Priestley after the Birmingham riots, expressing his condolences, replacing a book (by Adams) burned by the mob, and assuring Priestley of his welcome should he decide to emigrate to America. In the event, when the Priestleys arrived in New York in June 1794, Adams had left the city the day before, but left for Dr. Priestley a letter of welcome, urging him to consider settling in the Boston area.

Even by 1792, however, the seeds of discord between the two were beginning to appear. The youthful fires of revolution and democracy were perhaps burning a bit lower in Adams, and his faith in the wisdom of the people at large diminishing, when he wrote to Priestley, "Inquisitions and Despotisms are not alone in persecuting Philosophers. The people themselves, we see, are capable of persecuting a Priestly [sic], as another people formerly persecuted a Socrates".

In a letter the next year, he defended the monarchical system of government, anathema to Priestley. Personal relations between the two men nevertheless remained cordial for a time. As we have noted, Adams attended Priestley's "Discourses" in Philadelphia in 1796, and Priestley dedicated the published version to him. By the following year, however, differences especially in attitudes toward France resulted in an estrangement, and Adams did not attend Priestley's second set of sermons that January, nor a few months later did he reply to Priestley's request for assistance in arranging travel to France to look after some interests there. Indeed, under Adams's presidency (1797–1801), Priestley was in some peril of deportation or worse under the notorious Aliens and Sedition Acts, which Adams supported, but was spared by the President's not very complimentary intervention: "He is as weak as water, as unstable as Reuben, or the wind. His influence is not an atom in the world". His assessment had ameliorated somewhat when he wrote to Thomas Jefferson, years after Priestley's death, "Oh! that Priestley could live again!", and, in a more qualified vein to John Vaughan, "I never recollect Dr. Priestley, but with tenderness of Sentiment. Certainly one of the greatest Men in the World, and certainly one of the weakest".

Jefferson

Priestley's personal association with Thomas Jefferson began only in 1797, when Jefferson returned to Philadelphia from his country estate at Monticello in Virginia to assume the Vice-Presidency in John Adams's administration, and a post he may have valued even more highly, the presidency of the American Philosophical Society, for which Priestley had recommended him while turning down the honour for himself. Jefferson was certainly well acquainted with, and influenced by, Priestley's writings on philosophy, religion, and education, and apparently anxious to make his acquaintance. The two became well acquainted and saw one another frequently over the next few years. During this period Jefferson attended the Unitarian services in Philadelphia at which Priestley often preached.

Isaac Kramnick has summarized Jefferson's intellectual and philosophical debt to Priestley so well and so succinctly that it is difficult to paraphrase without plagiarizing. Jefferson's personal religious framework, his adoption of a materialistic view of the mind and soul, his commitment to disestablishment and the separation of church and state, his distrust of a strong central government, his focus on education as preparation for the future rather than reflection on the past, all owed much to Priestley's ideas and writings. Indeed, one of Jefferson's first acts on assuming the Presidency in March of 1801 was to write to Priestley, then seriously ill in Philadelphia, assuring him of his continued welcome in his adopted country, and at least implicitly relieving him of whatever anxiety he may have felt as to the possibility of legal sanctions under the preceding (Adams) administration. This is the letter that contains the oft-repeated encomium, "Yours is one of the few lives precious to mankind for the continuance of which every thinking man is solicitous". It has been suggested that this gracious and praiseful letter played a role in Priestley's recovery and survival for three more years.

Succeeding generations

A brief word about Priestley's children may be of interest, though time permits only a cursory recapitulation. His eldest child, Sarah, married William Finch, whose business career ended in disaster. She never left England, and died in 1803. His eldest son, Joseph, of whom we have said a good deal, returned to England after his father's death, and though he repeatedly expressed the intention of returning to America, never did. He died in 1833. The second son, William, remained in America, moving from Pennsylvania to Louisiana after having been accused, perhaps falsely, of attempting to poison his father. He became a successful planter, and like his brother Joseph left many and distinguished descendants. The youngest son, Henry, as mentioned previously, died in Pennsylvania as a young man. Figure 10 provides a skeletal summary of Priestley's descendants.

Concluding Remark

As is obvious from the foregoing, emigration to America did not completely free Priestley and his family from the controversy and obnoxious treatment that they had experienced in England. Priestley, however, never wavered in his basic admiration of his adopted country, but for me the most eloquent expression of the family's sentiments comes from the following paragraph, from a letter Joseph Junior wrote in 1820 from England to his son in Pennsylvania [the letter is preserved in the archives of Dickinson College]:

> I receive now and then a letter from H[***], but always full of complaints against the country. **I know very well that every thing is not perfect in America, & that there is very great room for improvement, not only in the Political institutions, but in the morals of the people of the country, & that on this side of the grave it must always be the case. The true way is to form a comparison between the condition of people in America, and that of any other people we hear of or are acquainted with. Where is there so great a portion of Civil and Religious liberty, where can a man so easily obtain the necessaries and comforts of life? In what country are the laws so well calculated to secure the freedom & happiness & rights of individuals?** Let any person dissatisfied with A[merica] consider what is going on here [*i.e.*, in Great Britain] where the constitution is set aside & the wealth & power of the government are arrayed against one individual, where

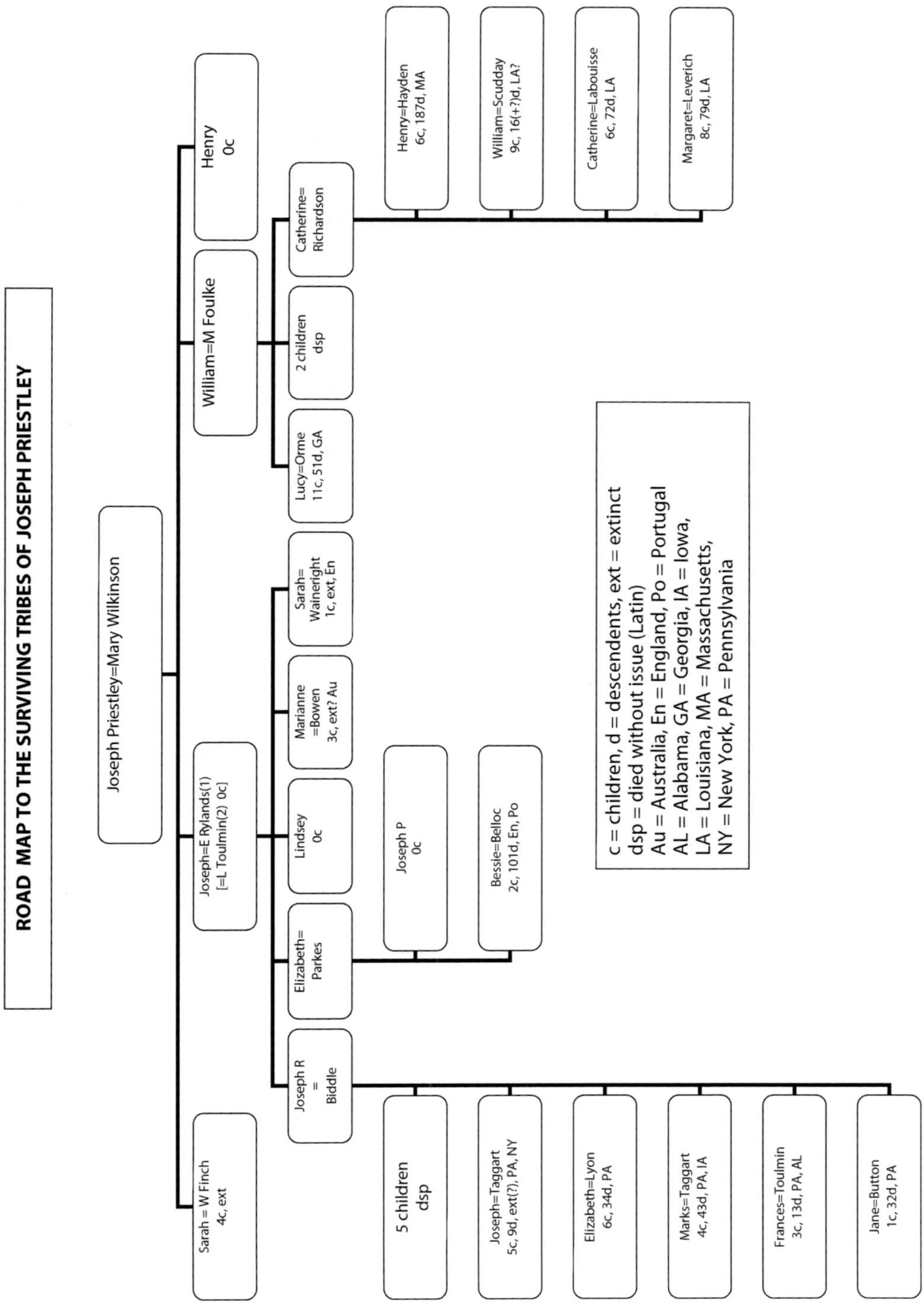

Fig. 10 Descendents of Joseph and Mary Priestley

hundreds are at this moment suffer[in]g imprisonment for the honest declaration of their opinions, where the will of the people is set at nought & where the object of the government seems to be anything else than the happiness of the people.

Some may see irony in these words today, and possibly even a role reversal, but setting aside the disgruntlement of the author with his current situation, they seem a fitting expression of the hopes and ideals of the country whose national day of celebration this is, and that Joseph Priestley sought when he came there.

A note on sources

I have chosen not to burden this account with detailed citations. Most of the facts cited are readily found in published materials, listed below. Jenny Graham's detailed monograph is an especially rich source, which I found invaluable.

Further reading

Jenny Graham, "Revolutionary in Exile/The emigration of Joseph Priestley to America/1794–1804. Priestley in America." Am. Phil. Soc. Trans. **85**, Pt. 2, p. 213, Philadelphia 1995.

F. W Gibbs, *Joseph Priestley: Adventurer in science and champion of truth*, Thos. Nelson, London, 1965

A. T. Schwartz and J. G. McEvoy, eds., *Motion toward perfection: The Achievement of Joseph Priestley.* Skinner House, Boston, 1990. Esp. contributions by Derek Davenport and Donald D'Elia on American period.

Robert E Schofield, *The Enlightenment of Joseph Priestley: A study of his life and work from 1733 to 1773*, The Pennsylvania State University Press, University Park, 1997.

______, *The Enlightened Joseph Priestley: A study of his life and work from 1773 to 1804*, The Pennsylvania State University Press, University Park, 2004.

______, ed., *A Scientific Autobiography of Joseph Priestley, 1733–1804*. The M. I. T. Press, Cambridge, Massachusetts, 1966.

Priestley Toulmin, "The Descendants of Joseph Priestley, LL. D., F. R. S", Northumberland County Historical Society Proc. **XXXII**, Pt. 2, p. 129. Sunbury, Pennsylvania 1994. (Part 1 of this volume contains several articles of interest relating to people and places associated with Priestley in America.)

SECTION 7

The Dickinson College Thomas Cooper Collection

JOSEPH PRIESTLEY AND SCIENTIFIC RESEARCH ON THE AMERICAN FRONTIER

Walter W. Woodward
Dickinson College, Pennsylvania

Walter W. Woodward

Benjamin Rush
CHARLES WILSON PEALE

Carlisle, Pennsylvania and Northumberland, Pennsylvania are both approximately 3500 miles from us, and 75 miles away from each other. At the time Joseph Priestley was in America though, these two towns shared a number of important qualities. Carlisle, like Northumberland, was a small town at the edge of the American frontier, on the boundary between land safely controlled by European settlers, and lands still fiercely fought for by the original Indian inhabitants.

Their locations made these two towns gateways to the west, which for a time led to Carlisle's being considered as a potential site for the new nation's capital.[1] Even then, people believed the future of the American experiment in republican government depended on its ability to expand successfully westward. For this reason, perhaps, Northumberland and Carlisle captured the attention, and commitment, of two of the most remarkable men of their age.

Joseph Priestley was one of these men,. Benjamin Rush was the other.

Rush is celebrated in America for many accomplishments. He was a signer of the Declaration of Independence, chief surgeon of the Continental army, a political and educational theorist, a chemical researcher and physician who helped found both the American Philosophical Society and America's first college of medicine.

Of all the major American figures with whom Joseph Priestley became well-acquainted, none was closer or more to his liking than Benjamin Rush. That is because both men shared a wide variety of interests and deep commitments to many of the same principles.

Both Priestley and Rush were intensely religious men, who believed they lived in a time of great spiritual importance, a time that might herald the onset of the millenium, Christ's thousand year reign on earth. Both were chemical

Detail from "MAP OF THE FRONTIERS OF THE NORTHERN COLONIES WITH THE BOUNDARY LINE ESTABLISHED BETWEEN THEM AND THE INDIANS AT THE TREATY HELD BY S.WILL JOHNSON AT FT. STANWIX IN NOVR. 1768", corrected and improved from the Evans map by Guy Johnson, Dept. Agt. of Ind. Affairs. E. B. O'Callaghan, A Documentary History of the State of New York Vol. I (Albany 1849). Carlisle is on the Susquehanna below the type "PENNSYLVANIA". Northampton, not named on the 1768 map, is along the Susquehanna above Carlisle, adjacent to the named Indian town of Shamokin

researchers who believed unflaggingly in the compatibility of the new sciences and scripture, who thought in fact that each could and would prove the truth of the other. Both were early, ardent public proponents of the abolition of slavery.

They were so whole-heartedly committed to the success of citizen-controlled republican government that both risked their lives to support that belief.

It should also be added that one of the bonds that undoubtedly brought them even closer was their contempt for a common enemy, the journalist William Cobbett, who, writing as 'Peter Porcupine', harangued both men's political views with a seemingly endless stream of satirical invective.[2]

Rush's commitment to millennialism and republican success was perhaps best expressed in his founding of Dickinson college – for which he developed a practical and spiritual curriculum designed to produce what he called the 'republican machines' who could assure the virtuous transfer of republican government and principles to the west.[3]

A mid-19th century photograph of "Old West" at Dickinson College, designed by Benjamin Latrobe, architect of the United States Capitol: DICKINSON CHRONICLES

Priestley too expressed his commitment to millennial republicanism by joining the group – including Thomas Cooper, Samuel Taylor Coleridge, and Robert Southey, who hoped to establish an American utopia on the Susquehanna.[4] In his Northumberland home, Priestley set up what he believed was the finest experimental laboratory in America. Like Benjamin Rush, Priestley hoped to found a college in the American woods, and worked assiduously but without success to find backers for his venture.[5] Though the college plan never came to fruition, Priestley was able to continue his scientific research agenda at his frontier laboratory in Northampton right up to his death in 1804.

Given the similarity between the interests of Rush and Priestley, there is something almost providential about the fact that much of Priestley's laboratory equipment made its way to Dickinson College several years after Priestley's death in 1804. Although the funds used to purchase the equipment

Portrait of Thomas Cooper, showing some of the chemical glassware from Priestley's collection

came from an account the trustees had given to Rush to procure scientific apparatus for the college, the agent most responsible for their transfer to the school was Thomas Cooper, the fellow radical republican, Unitarian, and chemist who had been a close friend of Priestley's since the Birmingham days, and with whom Priestley had settled in Northumberland.

Cooper, who had resided in Priestley's Northumberland home in the years before and following Priestley's death, was elected Professor of Chemistry and Mineralogy at Dickinson in the summer of 1811. This was the same year that Joseph Priestley, Junior, tired of the isolation and financial strains of rural Pennsylvania life, left Northumberland for Philadelphia and then England.

At Cooper's suggestion, and with Rush's assistance, the College paid Priestley, Junior, $530 dollars for a collection of Priestley's scientific apparatus including a telescope, Priestley's burning lens, and two orrerys.[6] Some of the Priestley equipment is seen below as it was photographed in the college science laboratory in 1875 by Charles Francis Himes.

Cooper, who was at odds with the College President even before he arrived, only remained with the College for four years before moving on to the University of Pennsylvania and then on to the Presidency of the College of South Carolina. Priestley's scientific apparatus has remained at Dickinson for nearly 200 years, and today his burning glass is the single most valued piece in the college's special collections.[7] The apparatus was in use at the college as

Priestley's scientific apparatus acquired by Dickinson College. The burning glass is in the middle.

late as 1880, as the side photograph taken by Charles Francis Heim reveals, but since then it has been retired to a place of honour in the College Archives.

The burning glass Priestley used in America was significantly larger than the one with which he discovered oxygen in 1774. It is made up of two lenses, one 16 inches in diameter, the other 7 inches in diameter. At its centre, the larger lens is 5 inches thick and the distance between the two lenses in the frame is 16 inches, the focal point of the larger lens. Like a modern telescope stand, the frame is placed on the stand in a way that allows it the user to move it both horizontally and to change its declination. By adjusting the angle between the sun's rays and the burning glass, heat could be focused on objects within closed glass tubes, a very useful tool to use when one is attempting to scientifically answer a question such as "Why do things burn?"

Because his explanation is, in my opinion, extraordinarily clear, this article repeats Charles Himes's 19th century description of 18th century theories of combustion. They help explain why the burning glass, rather than burning charcoal, was such an important instrument in Priestley's scientific experimentation.

> Why does charcoal, or wood, or anything burn? What is taking place to produce the light and the heat? The philosophers of that day said: When charcoal burns something is escaping from it, and that escape, somehow occasions the heat and light. That something, by common consent, they called phlogiston. The more rapid the escape of phlogiston, the more vigorous and combustible it was.
>
> But this phlogiston was a purely hypothetical something. It had never been isolated. But in all investigations it had to be reckoned with and carefully watched.

This, Heims went on to say, was what made charcoal fires a problem for experiments focused on finding phlogiston. Since everyone agreed that charcoal was very high in the substance, the phlogiston that escaped from it during burning might wreak havoc with any experiment's results.

A burning lens, on the other hand, offered real advantages, at least in theory. As Heims describes it, scientists assumed that sunlight was, or at least might be, free from phlogiston; *so they made large burning lenses, to make use of the heat of the sun.*

The burning glass made all the difference for Priestley. As he wrote to Benjamin Franklin, "Having at last got sunshine, I am busy prosecuting the experiments about which I wrote you, &c". And it was by means of a Burning Lens that Priestley discovered oxygen. When he focused the glass on a calx of mercury, he noticed that air was released. Capturing this air in a tube he tested its ability to combust and found that it made tapers burn more brightly. The reason seemed clear. This new air had less phlogiston mixed with it than atmospheric air. So it

With the burning glass, Priestley could focus the sun's heat on materials placed in tubes surrounded by water. This helped to keep the tests free from contamination.

was greedy for phlogiston and took it more rapidly out of combustibles. This surprising discovery Priestley called 'dephlogisticated air'.

Though Lavoisier would subsequently use gravimetrics to reconceptualize Priestley's discovery, naming it 'oxygen', Priestley had made a revolutionary finding with his burning glass, and he would repeatedly do so, ultimately discovering nine gases, more than any single individual before or since.

Priestley's burning glass represents an important moment of transition in the history of science The burning glass at Dickinson College is significantly larger and more powerful than the one with which Priestley discovered oxygen, and was one of the reasons Priestley was so sure of the superiority of his American laboratory.

The following 1975 picture of the astronomer Carl Sagan, seen peering through the glass the year he became the 23rd recipient of Dickinson College's annual Priestley Award, provides a useful reference for the burning glass's impressive size.[8] Other Priestley instruments at Dickinson include a 5 inch Gregorian reflecting telescope, still in excellent condition, manufactured by W. & S. Jones, London's leading maker of scientific instruments in the late 18th and early 19th centuries.

A four foot long achromatic refracting telescope made by John Dollond, celebrated member of the Dollond family which made fine telescopes for five generations, is also part of the Priestley collection.

Dickinson College also holds several items of Priestley's chemical glassware, and a somewhat ominous looking, but essential Priestley instrument known as a eudiometer, used to test the volumetric measurement of gases, from which the purity of the gas could be determined.

One of the most interesting items in the collection is Priestley's air gun, both because of its rarity, and the moment in which it appeared on the American scene, is an instrument that dramatically demonstrates the power of compressed air. Air guns, even if we disqualify the lung-powered blow-gun as a candidate,

Carl Sagan peering through the Dickinson College burning glass

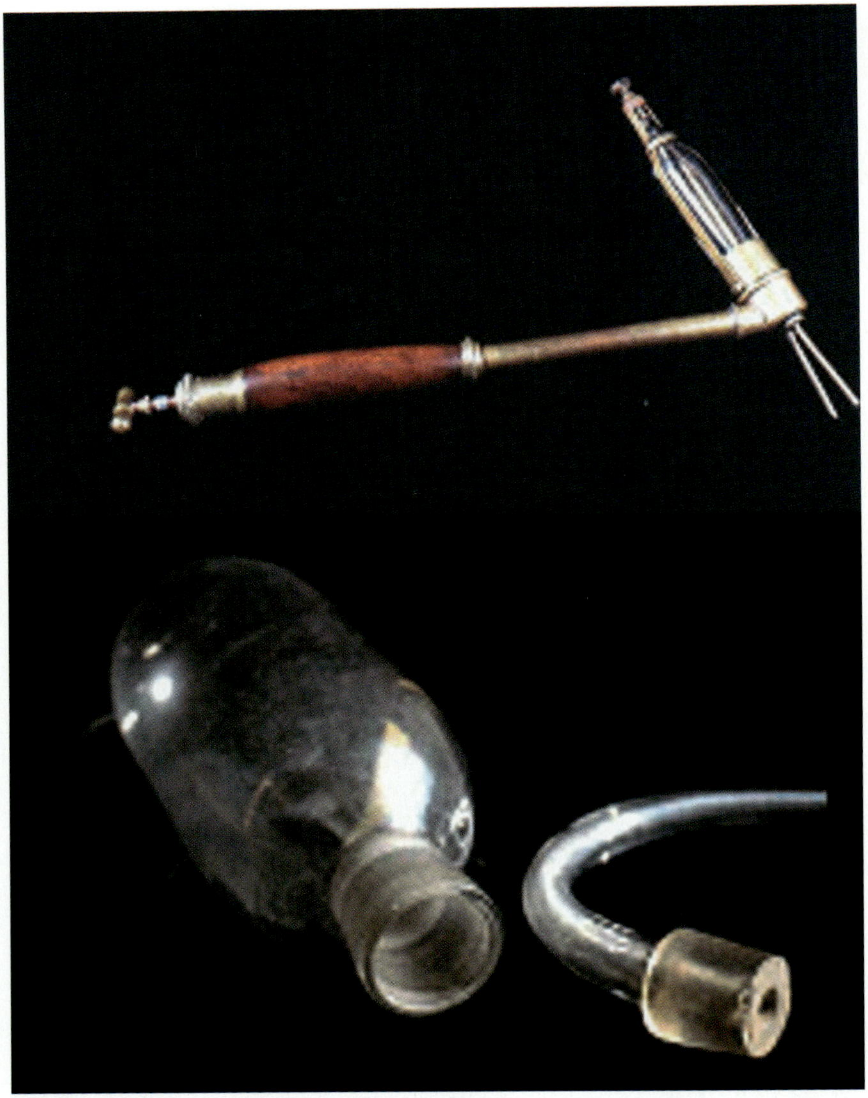

Eudiometer (*top*) and glassware

have a surprisingly long history. The oldest surviving mechanical air gun is a sixteenth-century bellows gun from Sweden. Around 1600, a special pneumatic pump air gun was made for France's King Henry IV. By 1800, air guns were among the most accurate large bore rifles, but the craftsmanship involved in their manufacture put them out of the range of all but the most elite. According to the historian Robert Beeman, all of these early air-powered weapons relied on using a hand-powered pump, like a bicycle pump, to compress air into a strong valved reservoir that was sometimes part of, but usually just attached to, the weapon.[9] Charging an air gun's reservoir took hundreds of strokes, and the resulting air pressure, under 600 psi, proved in most cases to be quite effective.

In Priestley's air gun, the reservoir for the compressed air was a 3½ inch wrought iron globe that could be screwed onto the firing mechanism. A short burst of air was released into the barrel when the trigger was pulled and the shot was propelled with force that made it effective for up to 150 yards. From Priestley's perspective, the major advantage of the air gun may have been its novelty, and the fact that it demonstrated another powerful utilitarian application of the air that figured in so much of his research. For hunters or soldiers, however, the potential advantages of air-powered weaponry were significant. Because the reservoir held a large volume of compressed air, air guns could be fired numerous times without repeating the cumbersome gunpowder and ball reloading process. With a gun like Priestley's, it would have been possible to have several reservoirs pre-filled and ready to replace each other for lengthy continuous firing.

Some air guns had a magazine that held multiple balls. Like a pin-ball machine, a ball was released into the barrel after every firing, making such air guns semi-automatic weapons. Another potential advantage of air guns was that they didn't make as much noise as gunpowder-fired weapons, nor did they have a firing pan or barrel flash that might give away one's location. The absence of smoke from the fired air gun also kept the line of fire visible to the shooter. These advantages – rapid firing capability, relative silence, and concealment of the firer – helped convince Meriwether Lewis and to take an air gun with him on his and William Clark's historic 'Voyage of Discovery' into the American West in 1804. Perhaps Lewis procured the air gun, out of concern that his men might run short of gunpowder during the multi-year journey. He may have planned on using the air gun as an alternate form of firepower.

He needn't have been concerned though, for the space-conscious Lewis had displayed remarkable ingenuity in planning. To protect the gunpowder he

The Priestley air gun. The spherical compressed air resevoir is at the top of the gun. Below the gun is the air pump used to recharge the (detached) reservoir.

carried for the regular firearms, Lewis had ordered 52 watertight canisters made of lead, each filled with four pounds of gunpowder. Once a canister was opened, it could be melted down to produce exactly the number of bullets four pounds of gunpowder could fire. As a result of this forethought and efficiency, The Corps of Discovery arrived back in St. Louis at the end of the three-year journey with much gunpowder to spare. But Lewis's air gun, surprisingly, was still one of the most indispensable pieces of technology carried on the trip west, despite the fact that it (and the voyagers) got off to a very unimpressive start. Lewis recorded an incident involving the air gun on the first page of his historic journal.

> Arrived at Bruno's Island, 3 miles below Pittsburgh ... went on shore and being invited by some of the gentlemen present to try my airgun which I had purchased brought it on shore charged it and fired myself seven times – fifty five yards with pretty good success; after which Mr. Blaze Cenas being unacquainted with the management of the gun suffered her to discharge herself accidentally, the ball passed through the hat of a woman about 40 yards distance cutting her temple about the fourth the diameter of the ball. She fell instantly and the blood gushing from her temple we were all in the greatest consternation supposed she was dead [but] in a minute she revived to our inexpressible satisfaction, and by examination we found this wound to be by no means mortal or even dangerous.[10]

Despite this rather inauspicious beginning, the curiosity about the air gun shown by the frontier people seems to have given Clark an idea that would help him use this technological marvel to best advantage. He began to employ demonstrations of the air gun – as the grand finale of a technological *tour de force* – in virtually every significant encounter with Indians on the journey west. Such 'shock and awe' displays are recorded in Lewis's journal 19 different times. The phrase that most frequently appears in conjunction with the air gun demonstrations is that the native peoples were 'much astonished' at its power.

During a meeting with a band of Sioux Indians, Lewis wrote "My air gun ... astonishes them very much, they cannot comprehend its shooting so often and without powder, and think that it is 'great medicine', which comprehends everything that to them is incomprehensible".[11]

During another meeting between the explorers and indigenous leaders at the continental divide, Lewis wrote: "Every article about us appeared to excite astonishment in their minds, the appearance of the men, their arms, the canoes, our manner of working them, the black man York, and the sagacity of my dog were equally objects of admiration. I also shot my air gun which was so perfectly incomprehensible that they immediately denominated it 'the great medicine'."[12]

According to Joseph Whitehouse, one of the men on the expedition, Indians had some help in deciding that the air gun had spiritual powers. Whitehouse reported the source of the claims to be Lewis himself, "Capt. Lewis shot his air gun and told them that there was medicine in her and that she would do great execution, they were all amazed at the curiosity, and as soon as he had shot a few times they all ran hastily to see the ball holes in the tree. They shouted aloud at the site of the execution she would do &c".[13] One of the earliest attempts by Americans to use advanced technology to assert authority, and in the case of presenting themselves to native peoples who vastly outnumbered them, it may have helped them earn respect and save lives – most notably their own.

Historians of firearms debate almost every aspect about Lewis and Clark's airgun. Who made it? Was it a single ball rifle, or did it have an automatic reloading mechanism. Equally, they wonder who influenced Lewis to take along an air gun in the first place. Some argue that Coleman Sellers, brother-in-law of the artist Charles Rembrandt Peale, was known to have visited Lewis and may have recommended the purchase, because Sellers was known to have had an airgun.[14]

But the air gun owner – better known to Lewis and Thomas Jefferson, the man for whom Lewis worked and who got Congress to fund the expedition – was Joseph Priestley. Jefferson deeply admired Priestley. Immediately after Jefferson was inaugurated President in 1801 he wrote to Priestley saying that now that he (Jefferson) was in office "science & honesty are replaced on their high ground; and you, my dear Sir, as their great apostle, are on its pinnacle". Lewis, as Jefferson's confidante and amanuensis, would certainly have known of the esteem in which the President held the sage of Northumberland, Pennsylvania and, as Jefferson and Priestley were in contact in 1803 just as the Voyage of Discovery was being planned, it is possible that the idea of taking the specially interesting technology of the air gun westward originated in some unrecorded or as yet unfound communication with Priestley.

Mrs. Temple Fay

Priestley's burning glass, telescope, air gun, and chemical glassware – instructive as they are about the nature and value of research and technology on the Early American frontier – represent only one part of the remarkable Priestley collection housed at Dickinson College.

In 1965, Mrs. Temple Fay, a fifth-generation Priestley descendant and widow of a renowned Philadelphia physician, augmented the College's scientific collection with a remarkable collection of Priestley material collected by her father, Joseph Priestley Button.

Included in this extensive collection are 84 contemporaneous books and pamphlets by Joseph Priestley. There are 92 works about Priestley, including the most inflammatory and polemical attacks on him written during his lifetime.

There are 148 family letters and documents extending from Joseph Priestley's forebears, through his own career, and on to succeeding generations.

"The friends of the people", cartoon by Isaac Cruikshank. Thomas Paine (*right*) and Joseph Priestley (*left*) with a 'malevolent spirit' in the middle.

(*opposite*) Below the Peale portrait of Priestley at Dickinson College is the Longman and Broderip barrel organ. On top is a pin-studded cylinder used to produce the music.

In addition there are newspaper clippings, photographs, and ephemera of special interest to those interested in the way in which the history of science is remembered and interpreted through time.

Along with this significant collection of printed matter, the college received 54 engraved portraits and caricatures of Priestley, many of which have interesting stories of their own to tell.

One of the most important of these images is the 1801 painting of Priestley by the celebrated American artist Rembrandt Peale. Peale made two copies of this painting, each for members of the American Philosophical Society. The College's copy is the one painted for Thomas Vaughan, Priestley's close friend both before and after his emigration, and librarian of the Society. Unlike the other paintings, Peale in this version included in the background volumes of Priestley's works.

Priestley figurine

At Dickinson, the Peale portrait hangs above another unique Priestley artifact, the Longman and Broderip barrel organ that Priestley used in his American home. The hand-cranked instrument worked much like the player pianos we are more familiar with. The crank operated a small bellows and rotated a cylinder – the barrel – thick with carefully placed protruding pins. The rotating pins opened the organ valves, releasing the pressurized air to play the tunes. Each barrel played 10 tunes. Priestley's organ is equipped with three barrels: one of classical music, including selections from Handel's "Water Music"; a second played hymns and songs, undoubtedly useful for a minister; and the third played popular airs, several of Scottish origin.

One of these songs, the "Birks of Endermay," was particularly moving to Priestley's friend Benjamin Rush.

As a medical student in Edinburgh in the 1760s, Rush had fallen in love with Lady Jean Leslie, daughter of the Earl of Leven. They had parted when he returned to America, and both married others. During the American War for Independence, however, Rush was grief-stricken to learn that Jenny's Brother, Captain William Leslie, had been killed at the Battle of Princeton by American troops on December 26th, 1776. He saw to his burial with full military honours, and after the war, paid to have a tombstone placed at the site, and sent Jenny an image done in hairwork of an imagined scene of Rush showing Jenny her brother's tombstone.

It closes with an invitation to the reader to journey back to the Pennsylvania frontier some two hundred years ago and look in on an imagined scene: of Benjamin Rush and his friend Joseph Priestley, sitting by the organ and singing together "The Birks of Endermay".

> The smiling Morn, the breathing Spring,
> Invite the tuneful Birds to sing:
> And while they warble from each Spray,
> Love melts the universal Lay.
> Let us, *Amanda*, timely wise,
> Like them improve the Hour that flies;
> And in soft Raptures waste the day,
> Among the Birks of *Endermay.*
>
> For soon the Winter of the Year,
> And Age, Life's Winter, will appear:
> At this, thy living Bloom will fade;
> As that will strip the verdant Shade.
> Our Taste of Pleasure then is o'er;
> The feather's Songsters love no more:
> And when they droop, and we decay,
> Adieu the Birks of *Endermay*.

One would be thinking of a land and friends far away; the other of love lost and the horrible cost of war, and both perhaps, of the price they had paid for liberty. Neither could ever have imagined how deeply their legacies would be intertwined in the centuries to come, or that a college founded by the one would become home to so many of the precious artifacts of the other.

NOTES

1. Joseph J. Ellis, Founding Brothers. *The Revolutionary Generation*, New York, 2000, p. 70.
2. William Cobbett, *Observations on the emigration of Dr. Joseph Priestley, and on the several addresses delivered to him, on his arrival at New-York*, Philadelphia: Printed by Thomas Bradford no. 8 South-Front Street, 1796; Benjamin Rush, William Cobbett, Thomas Carpenter, and Pennsylvania. Supreme Court. *A report of an action for a libel brought by Dr. Benjamin Rush, against William Cobbett, in the Supreme Court of Pennsylvania, December term, 1799, for certain defamatory publications in a news-paper, entitled Porcupine's gazette, of which the said William Cobbett was editor*, Philadelphia: Printed by W. W. Woodward no. 17 Chesnut Street, 1800.
3. Benjamin Rush, "Of the Mode of Education Proper in a Republic" in *Essays, Literary, Moral & Philosophical*, Philadelphia, 1798, p. 14, passim.
4. Edward Dowden, *Southey*, Harper & brothers, New York, 1880, p. 35.
5. Joseph Priestley, Thomas Cooper, and William Christie, *Memoirs of Dr. Joseph Priestley, to the year 1795*, Printed for J. Johnson, London, 1806, passim.
6. Charles Coleman Sellers, *Dickinson College: A History*, Middletown, CT, 2001, p. 153. Online version at http://chronicles. dickinson. edu/histories/sellers/index.html
7. Dumas Malone, *The public life of Thomas Cooper, 1783–1839*, University of South Carolina Press, Columbia, 1961.
8. The Joseph Priestley award is given annually by Dickinson College to a scientist who has made an important contribution to human knowledge. To date, fourteen Nobel Prize Recipients have been among the recipients of this award.
9. Robert Beeman and Ulrich Eichstädt, "Wohner der Wind weht" V*ISIER*, *Das internationale Waffenmagazin*, April 2002, vol. 4, pp. 134–142.
10. The Journals of the Lewis and Clark Expedition, Meriwether Lewis, August 30th, 1803, University of Nebraska Press, Stewart (AAGS), Chatters, Halsey, Wolff, pp. 131–32. http://lewisandclarkjournals.unl.edu/1803–08–30.html.
11. Meriwether Lewis, January 24th, 1806 in Gary E. Moulton, ed., *The Journals of the Lewis and Clark Expedition*, 11 volumes, Univ. of Nebraska Press, Lincoln, 1983–).
12. Meriwether Lewis, August 17th, 1805, ibid.
13. Gary E. Moulton, *The Journals of the Lewis & Clark Expedition: The Journals of Joseph Whitehouse, May 14th, 1804–April 2nd, 1806*, Journals of the Lewis & Clark Expedition Series, vol. 11, Lincoln NE, 2003.

14. Stewart Henry, Jr."The American Air Gun School of 1800 (with Corollary Verification of the Lewis and Clark Air Rifle Maker) in *Monthly Bugle*, February 1977, pp. 2–7; Halsey Ashley, Jr."The Air Gun of Lewis and Clark." *American Rifleman*, August 1984, pp. 36–37, 80–82; Eldon G. Wolff, *Air Guns*, Milwaukee: Milwaukee Public Museum, 1958; Robert D. Beeman, "Proceeding On To The Lewis and Clark Airgun – II", http://www.beemans.net/Lewis%20&%20Clark%20Airgun.htm

SECTION 8

BOC Priestley Bicentennial Symposium

BIRTH OF CHEMICAL AND BIOLOGICAL SCIENCES

The symposium, held at the University of Huddersfield on 4th–6th July, 2004, was attended by about 70 persons. It formed the first major meeting of the Priestley Society since its inauguration at Oakwell Hall to celebrate the life of Joseph Priestley in the year of the 200th anniversary of his death. It was also appropriate that the first day of the meeting was American Independence Day because of Priestley's well-established support for American independence from British rule in the late eighteenth century. The meeting comprised invited lectures on a variety of subjects related to the life and interests of Priestley, together with an associated variety of exhibits.

Grateful acknowledgement is hereby made to the following for financial aid making the Symposium possible: BOC Group; Rohm and Haas Company; Bristol and District, East Midlands, Huddersfield, and Central Yorkshire Sections, and North East Division of the Royal Society of Chemistry.

The following pages present the lectures given during the three days of the meeting.

The Symposium participants

Top Table: Mr. Gary Williams (BOC), Sir John Rowlinson, Dr. Priestley Toulmin, Prof. Fenn-Arthur (Deputy VC, University of Huddersfield), Cllr. Mary Harkins (Mayor of Kirklees), Dr. Stephen Bartle (Vice-Chairman, Priestley Society), Mrs. Toulmin, Dr. Norman Rose

Address by Dr. Stephen Bartle, Vice-Chairman Priestley Society, Master of Ceremonies

Ladies and Gentlemen, we are gathered here to celebrate the life and legacy of Dr Joseph Priestley.

It gives me great pleasure to welcome the Worshipful, the Mayor of Kirklees, Councillor Mary Harkin who has a great interest in Joseph Priestley.

The Mayor has graciously consented to open our celebrations with an address of welcome to our guests, Madam Mayor the rostrum is yours!

(*The Mayor gave a speech of welcome*)

Thank you, Madam Mayor for your warm and kind sentiments

And now the Deputy Vice-Chancellor of the University, Professor Fenn-Arthur will reply to the Mayor.

(*The Deputy Vice-Chancellor gave a welcoming reply*)

Finally I would like to express our thanks to Mr. Gary Williams of BOC for its generous support of this prestigious tribute to Dr. Priestley.

(*Mr Williams gave a reply*)

Ladies and gentlemen, we will now say Grace

Almighty God, for what we are about to receive may we be truly thankful. Amen

And now, Madam Mayor, Ladies and Gentlemen, a few words about our hero.

The inspiration throughout the life of Dr. Joseph Priestley was his belief in the humanity of Jesus Christ, and the love and compassion that He showed for human beings throughout His ministry on earth.

Thus this remarkable man devoted his prodigious energy and intellect to improving the well being of mankind, and of the common people; for example, all his scientific work was ultimately directed to practical application.

Thus, he is now regarded as the *Father of Modern Chemistry* who identified nine gases and encouraged the study of the subject.

And by the same token the *Father of Modern Physics* in Electricity, Optics, Astronomy and Acoustics.

In *Botany* he discovered photosynthesis.

He promoted new ideas in education for the young, for adolescents and for women.

He inaugurated modern ideas towards commercial studies in languages and mathematics.

He also recommended vocational training and recreational activities for students.

In addition to his belief in the education of women he sought equal status for them in society with men.

He inspired hot air ballooning through M. Montgolfier; and he encouraged others in their studies, for example, James Watt in chemistry.

He was an historian who produced a monumental *Chart of Biography* to illustrate how civilizations developed and interacted across the world from the earliest times. This gave him an insight into the history of ideas and earned him the degree of Doctor of Laws from Edinburgh University; (this was not an honorary degree, as sometime claimed).

He pioneered Constitutional Reform, and laid emphasis on the responsibilities of government to serve the interests and happiness of the common people, thus he came to play a significant role in the formulation of the written constitutions of the United States and of France.

As far as England was concerned, he had his own agenda. Many of the reformers of Priestley's day performed acts of considerable charity and good works, but the one difference between their commendable efforts and those of Joseph Priestley, I would suggest, is that with his scientific analytical mind, instead of trying to treat the symptoms of a society failing the ordinary people, Priestley identified the very source of the problem and single-mindedly he attacked the very core of the disease, *i.e.*, the rapidly diminishing franchise (down to 2%). This was to cost him dear and he lost his home, his library and possibly the finest laboratory in the world, his life's work and full recognition of his achievements in Britain.

The official view from the King (to the Home Secretary) reads "As the mischief did occur, it was impossible not to feel pleased at its having fallen on Priestley rather than another, that he might feel the wickedness of the doctrines of democracy he was propagating"!

The profound irony of the situation came out in a few weeks, when Louis XVI accepted the new constitution of France, which gave votes on a generous ratio to taxpayers in a constituency unlike Britain, where Priestley pointed out that Yorkshire with nearly one million of souls, sends only two county members to Parliament, while Manchester with 60000 does not send one!

Finally, Burke said that Priestley had received honorary French citizenship in 1792 in return for defaming the British constitution, Priestley replied that this was a lie, and as Wilberforce received the citizenship amongst others at the same time as Priestley, why did not Burke also accuse him of the same calumny; Burke never replied.

Commenting on their cruel acts, Mrs Barbauld – a family friend – wrote in 1792

To thee the slander of a passing age
Imports not. scenes like these hold little space
In his large mind, whose ample stretch of thought
Grasps future periods – Well can'st thou afford
To give large credit for that debt of fame
Thy country owes thee. Calm thou can'st consign it
To the slow payment of that distant day,
If distant, when thy name, to freedom's join'd,
Shall meet the thanks of a regenerate land.

(Poems of Anna Laetitia Barbauld)

It seems that the land is not yet regenerate!

He opposed negro slavery and he promoted religious toleration, and universal civil rights for minorities.

He believed in peaceful intercourse between the nations, in commerce and in scientific knowledge.

He wrote many works on theology, firmly based on his remarkable knowledge of ancient languages.

In his scientific work Dr. Joseph Priestley aimed to establish a disciplined meticulously conducted and recorded scientific method in all his experiments.

Above all, our hero possessed an unlimited fund of Christian forgiveness towards his enemies.

Now let us draw a mental image of him from several who knew him ... "In his countenance, though you might discern the Philosopher, yet it beamed with so much simplicity and freedom as made him very easy to access". A Quaker lady said of him "A man of admirable simplicity, gentleness and kindness at heart, united with great acuteness of intellect".

Finally, a new breakthrough in Priestley research – Professor Leslie Woodcock believes that the theory of phlogiston, as interpreted and advocated by Priestley to his dying day, although long obsolete can be made compatible with modern ideas of thermodynamics!

His energy, good humour, patience and generosity of spirit, always sharing his knowledge and discoveries with others without inhibition, gives us the model of practical and divinely inspired Christianity. His highly organised life was full of fun, especially for young people for whom he always found the time to care.

Let us celebrate the exceptional life of Dr. Joseph Priestley with joy and enthusiasm, and an unquenchable optimism and faith in the Almighty.

Madam Mayor, Ladies and Gentlemen, let us stand to toast Joseph Priestley, the greatest Britain ever; Joseph Priestley of Birstall and Heckmondwike, to his life and eternal legacy; Joseph Priestley.

And now Madam Mayor, ladies and gentlemen, we have a very special treat – Mr. Ronald Blatchley a retired chemistry teacher, has spent almost half a life time studying and performing what can only be described as a virtual reincarnation of Dr. Joseph Priestley at Priestley House, Northumberland Pennsylvania. Ladies and Gentlemen, Dr. Joseph Priestley in person."

"Dr. Priestley we thank you most warmly for your honoured presence in our celebration of your life. Thank you from the bottom of our hearts."

"Priestley" comes to Huddersfield

Oxygen Limericks by Ron Blatchley

Joe Priestley, the cleric and sage
Helped found a new chemical age
With an air that he rated
Dephlogisticated
When it caused glowing embers to rage.

Said Joe to the Earl of Shelburne,
"In this air lots of stuff will pell-mell burn.
A taper that's lame
Will burst into flame
And a lump of brimstone will like hell burn."

He tried his new air on two mice
Seeking their expert advice.
When it did not cause death,
He himself took a breath
And thought that it felt really nice.

His breast felt peculiarly light
And he thought maybe someday folks might
Under medical care
Take a whiff of his air.
And today we all know he was right!

Royal Society of Chemistry

LOCAL SECTIONS

RSC Priestley Bicentennial Lecture

DISCOVERING THE FORCES WITHIN MATTER

John S. Rowlinson
University of Oxford

Sir John S. Rowlinson

Abstract

The question of what holds matter together has been asked throughout the ages. Newton and his followers attempted an explanation, based on forces acting between the particles of matter, but this ran into experimental, theoretical and metaphysical objections throughout the 18th century. Laplace brought it back to favour with his theory of capillarity early in the 19th century but it was not until the second half of that century that it became firmly established. An understanding of the origin of the forces had to await the quantum theory of the 1920s and the metaphysical objections to action at a distance were not resolved until 1946. The ebb and flow of theories in this field is used to illustrate the contingent nature of scientific advance.

Introduction

The study of the intermolecular forces that are responsible for the cohesion of matter now forms a recognized branch of physical chemistry that was much studied in the 20th century. In this paper I examine how our ideas on this subject have evolved from the beginning of the 18th century onwards and set out the different physical and metaphysical assumptions that have characterised this field over the years. It has not been a smooth evolution but one in which chance and fashion have played a big role.

The scientific revolution of the seventeenth century was marked by the rout of the 'occult forces' of scholastic philosophy. It was the firm belief of Descartes, Huygens, Leibniz and their followers that the physical world should be explicable in terms of mechanical effects. Thus Descartes explained the motion of the planets around the sun by postulating the existence of vortices of invisible matter that carried them on their elliptical tracks.[1] Newton, however, showed that an inverse-square force of attraction between the sun and the planets provided a better explanation, not only of their motions and that of the moon, but also of terrestrial gravitation and of the rise and fall of the tides. Such a force, acting at a distance, was repugnant to the Cartesians who accused Newton of re-introducing the occult forces that they had so recently banished. Indeed, Newton himself agreed that the idea of action at a distance was 'absurd' and accepted that "Gravity must be caused by an agent ... but whether this agent be material or immaterial, is a question I have left to the consideration of my readers".[2] He agreed with John Locke that it is "impossible to conceive that a body ... can operate where it is not".[3] All he insisted on was that, however it was caused, there was such a force and that it decreased as the inverse square of the separation of all material objects. Nevertheless such positivist reticence did not placate his Continental critics who were unwilling to allow that a force might exist if they could not assign an explicit cause to that force. There was therefore bound to be a hostile reception to Newton's extension of his ideas on attraction to explain the cohesion of matter by means of similar forces acting between the particles from which he assumed all bodies to be constructed.[4] This extension, from the planetary to the terrestrial, to the microscopic, arose from his firm belief in the consistency of the Creation, for, he said, "if Nature be most simple and fully consonant to herself she observes the same method in regulating the motions of smaller bodies which she doth in regulating those of the greater".[5] A law that held for planets and apples must surely hold, although possibly in a different form, for all particles of matter.

The early years of the 18th century was therefore a time of competing experiments and theories about cohesion, or what it was that held matter together. There were those who did not believe in a corpuscular constitution of matter; there were those who accepted the corpuscles but would not allow that they exerted forces on each other unless they had a mechanical explanation of how those forces acted; and there were the Newtonians who found the evidence for the forces to be compelling but admitted that they could not explain what caused them or how they acted.

Newton, Freind and Hauksbee

The distinction between what we now call physical and chemical forces was scarcely recognised at the start of the 18th century; both are described in the famous Query 31 that concluded the last editions of Newton's *Opticks*.[4] Before

Newton's ideas on attraction had appeared in print his followers in Oxford, John Keill and John Freind, had expounded them publicly. Freind gave a course of lectures in 1704 which he published five years later, probably in a revised form after the appearance in 1706 of the Latin edition of the *Opticks*. Here he tried to explain all chemistry in terms of different forces between microscopic particles of matter. He, like many other followers of Newton's philosophy, often abandoned their master's cautious positivism and alleged that it was clear that the forces between the particles acted at a distance. When his book was criticized in a review by Christian Wolff, an associate of Leibniz, Freind wrote, in obvious exasperation: "Such a principle of attraction they are pleas'd to call a figment, but how any thing shou'd be a figment, which really exists, is past comprehension".[6] Freind's insights included the observation that the forces responsible for cohesion must fall off more rapidly with distance than the inverse-square force of Newton's gravity. His attempts to interpret the ways in which different acids attacked different metals by means of algebraic formulae were, however, not a success. The world was not ready for quantitative physical chemistry for another 150 years.

But perhaps more influential than Freind's theorising was the experimental work carried out at the Royal Society early in the 18th century by Francis Hauksbee, during Newton's presidency, and clearly under his direction. Hauksbee's experiments covered a wide range of phenomena but it is those on capillarity that contributed most to the debate on the cohesion of matter. The rise of liquids in capillary tubes, in apparent defiance of the laws of hydrostatics, intrigued many natural philosophers throughout the 18th century and this field became one of the places where the Newtonians and the Cartesians joined battle. Hauksbee established first that the rise of a liquid in a tube is the same in a vacuum as in air, thus disproving a much earlier notion of Newton and others that the rise is due to a reduction of air pressure within the tube. He then showed that a liquid rose also between parallel plates of glass, and either he or James Jurin was probably responsible for the experiments behind Newton's observations in Query 31 that this rise is inversely proportion to the spacing of the plates and that it is only half of that in a tube of a diameter equal to the spacing. These experiments established the quantitative laws that govern capillary rise. Perhaps the most ingenious, and potentially the most important, of the experiments that Hauksbee made, although one that Newton did not discuss, was to show that the rise is independent of the thickness of the wall of the glass tube and dependent only on the internal diameter. To this observation Jurin added that it is only the internal diameter at the height of the meniscus that is relevant, not that at lower points, so that a tube narrower at the top than the bottom can support a greater weight of water in suspension than one of uniform bore. If the rise is independent of the thickness of the walls of the tube then, Hauksbee argued, "the attractive power of small particles of matter acts only on such corpuscles as are in contact with them, or remov'd at infinitely small distance from them".[7] When, however, he tried to

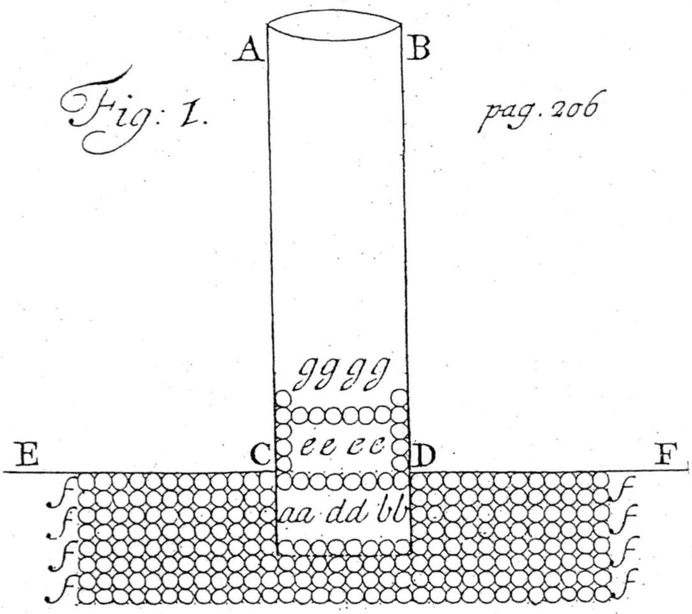

Figure 1. Hauksbee's picture of the rise of particles of water in a capillary tube (1709).

explain how the horizontal attraction of the glass walls induces a vertical rise of the water his argument became less coherent and required that the forces can act out to a distance of the internal radius of the tube (Figure 1).

Brook Taylor, a Secretary at the Royal Society, collaborated with Hauksbee in trying to discover the laws of magnetic attraction, but without success with the crude lodestones at their disposal. To one of their papers, however, Taylor attached a note on an experiment that was to be much-repeated later in the century in which he measured the force needed to detach a flat piece of wood from the surface of water in order to determine the attraction between wood and water.[8]

Boscovich and Priestley

Repulsive forces between the particles in collision played a lesser role than the attractive in the Newton's scheme. He had speculated that gas pressure arose from a repulsion between static neighbouring particles of air that was inversely proportional to their separation, but this idea, although often tacitly accepted, aroused little discussion since it seemed to lead nowhere. His view of the composition of the fundamental particles was quite complex but as long as he supposed them to occupy space, and so be impenetrable, he did not need to speculate further about the existence of repulsive forces. Stephen Hales and John Desaguliers were the first Newtonians to discuss seriously the existence of such forces when they sought, respectively, to explain the expulsion of 'airs' from fermenting material and the elasticity of solids.[9] A more radical approach was taken in 1758 by Roger Boscovich, a Jesuit priest from Ragusa [Dubrovnik]. He dispensed with particles with hard cores and proposed instead that the particles were massy points that repelled each other on close approach with a force that varied steeply but continuously with their separation. He went further and supposed that there were alternating zones of attraction and repulsion as the particles moved farther apart, leading to positions where the force was zero and so where there could be particles held together by physical or chemical attractions balanced by the repulsion (Figure 2). He wrote that completely hard or incompressible bodies "never occur in Nature, still less can they occur in my theory".[10] This was a bold statement at the time since the compressibility of liquids and solids was still a matter of doubt until 1762 when John Canton first measured, with excellent accuracy, the compressibility of water.[11]

Boscovich's views attracted surprisingly little discussion on the Continent, possibly because he quarrelled with many of the leading mathematicians of the day, but they were sympathetically received by Joseph Priestley in 1772, who wrote that his Yorkshire neighbour, the astronomer John Michell, had held similar views to those of Boscovich.[12] (Michell had earlier produced the first artificial magnets and had provided strong evidence that the magnetic poles exerted mutual forces that, like gravity, conformed to the inverse square law.[13]) Priestley had previously believed that the constituent particles of matter were hard bodies, but was persuaded by 'Father Boscovich and Mr Michell' to abandon that view and to accept that they were kept apart by a continuously varying force. At the age

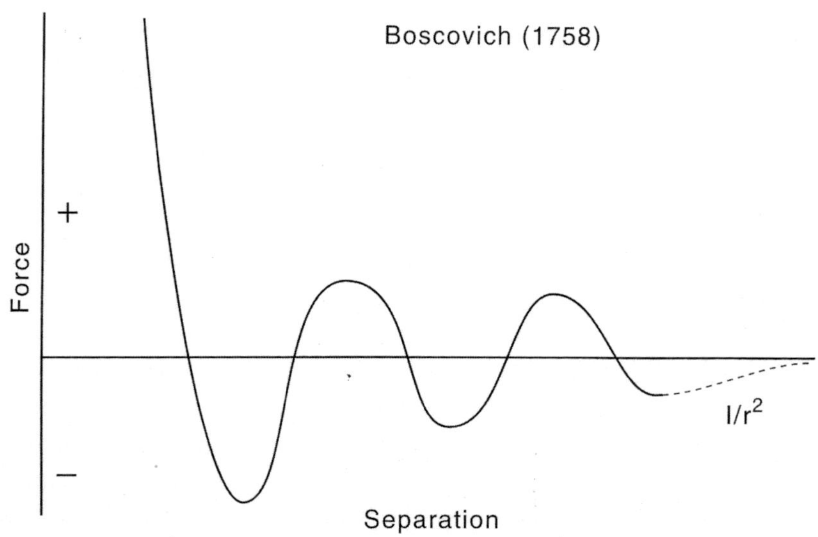

Figure 2. Boscovich's representation of the force between a pair of particles (1758)

of 22 he had read the views of the Rev. John Rowning, written in 1738, which were very similar to those of Boscovich, but he had presumably not then taken them seriously.[14] In his natural philosophy Priestley broadly followed Newton's 'physicalist' account of chemistry although he differed from him in his theological interpretation of natural philosophy.[15] His account of Boscovich's work in 1772 may mark its first appearance in English. Boscovich's model also attracted attention from the Scots philosophers, although Priestley disagreed with them also on matters of theology.[15, 16] In the 19th century the British physicists who promoted field theories found the model more congenial than the hard particles of traditional atomic theory; Lord Kelvin, for example, was still using the expression 'Boscovichianism' in the 20th century.[17]

'A secondary order of men'[18]

After an encouraging start at the hands of Newton and his followers in the early years, the rest of the 18th century was marked by the relative neglect of the subject of cohesion and capillarity, and those who took up this field were generally not the leaders of the day. In part this was due to the inherent difficulties of these subjects and in part due to competition from the more exciting fields of electricity, magnetism and heat, where real advances were being made. The undeniable success of Newton's gravitational theory meant that the astronomers slowly abandoned their opposition to its apparent invocation of action at a distance. This acceptance led during the century to a willingness to apply Newton's ideas also to the less easily studied forces between particles of matter. When it was found that electric charges and magnetic poles also followed the inverse-square law then apparent action at a distance was admitted, reluctantly and often only implicitly, into the realm of natural philosophy. It was agreed that it was unprofitable to ask too closely exactly how such forces were transmitted. Those who accepted Newton's doctrine of inter-particle forces were to be proved right but they could not translate their good measurements into a coherent explanation of how the supposed forces brought about the observed phenomena.

The first French and Italian workers on capillarity in the opening decades of the 18th century made many miscellaneous experiments that led nowhere. They were then still committed to seeking Cartesian explanations but could not agree what these should be. Their experiments were less useful than those of the Newtonians, as is often the case of experiments without a clear objective when compared with those guided by a theoretical idea, even if that idea is not always entirely sound. The Dutch were more readily converted to Newton's views and some of the best measurements of capillary rise in the 18th century were made by Pieter van Musschenbroek, first at Utrecht and later at Leiden. Towards the end of the century many of those in this field, dismissed by John Leslie as a 'secondary order of men',[18] repeated 'Dr Taylor's experiment' of measuring the force needed to lift a flat plate from the surface of a liquid, but again could give no explanation of the effects that they had measured. It is easy to see in retrospect that the key idea that was missing from all the many discussions of capillarity and similar phenomena was that of surface tension. It was known that flies could walk on water and that, with care, a needle could be made to float on the surface, but these observations were not, for a long time, translated into the perception that there was a tension associated with the surface. There was early talk of stretched threads at the surface, possibly because this was the standard model used to interpret the elasticity of solids, but the first clear idea of a surface tension that acted uniformly in all directions did not appear until a paper from the French mathematician Gaspard Monge in 1789.[19] He, however, never followed up his promising start on this field.

Young and Laplace

The indisputable success of Newton's gravitational theory meant that during the 18th century the idea of forces that *apparently* acted at a distance came to be accepted, not only for the phenomenon of capillarity, but also for 'Dr Taylor's experiment' and for the forces that led to chemical reactions. But this sometimes grudging acceptance did not solve the important problems. A French Newtonian wrote in 1754: "It is not sufficient to say, in a vague way, that attraction is the cause of the suspension of water in capillary tubes; one must explain how the attraction acts, and there is the difficulty".[20]

A solution to this problem came in the opening years of the 19th century in papers from Thomas Young (1805)[21] and Pierre-Simon de Laplace (1806).[22] They addressed the same problem and arrived at concordant results, but their two papers are of very different character. In reading Young we are reading the natural philosophy of the 18th century, in reading Laplace we find the theoretical physics of the 19th, securely based on the mathematical teaching of the new institutions of the French Revolution. Young deserves the credit for the first publication of a solution to the problem of capillarity but Laplace soon went much further than Young. His style is so much more to the modern taste that it is easier to restrict the discussion to his results.

Laplace starts from three assumptions; first, that the inter-particle forces, like gravitational forces, are transmitted through matter without any obstruction; second, that for each liquid and solid (water and glass, for example) there is a fixed angle of contact that is determined, in a way that need not be specified explicitly, by the strength of the water-water and water-glass attractions; and third, that the range of the forces is short by comparison with any relevant macroscopic distance, such as the radius of the capillary tube. The first two assumptions both he and Young seem to regard as obvious, but for the third he relies on what he believed to be Hauksbee's conclusion from his experiment with tubes of different wall-thickness. He encapsulates this assumption in a phrase that he uses repeatedly; the forces "are sensible only at insensible distances". With these three assumptions he is able to show that the stress arising from the inter-particle forces, or what is now usually called the internal pressure, P, within a drop of liquid of radius R, takes the form:

$$P = K + H/R, \qquad (1)$$

where

$$K = 2\pi \int \Psi(r)dr \qquad \text{and} \qquad H = 2\pi \int r\,\Psi(r)\,dr. \qquad (2)$$

Here $\Psi(r)$ is the second moment of the attractive inter-particle force at a separation r, that is, the integral over all separations of $r^2 f(r)$ where f is the force. This expression for the internal pressure clearly requires that the strength of the force falls off with separation sufficiently rapidly for the integrals to converge at infinity. He left undetermined what happened at the lower limit of the integrals, when the separation becomes zero.

Equation (1) is now called Laplace's equation; it tells us that the second term is the difference of pressure within a liquid with a planar surface ($R = \infty$) and in one with a curved surface of radius R. The fixed angle of contact of liquid and glass means that the meniscus of a liquid in a capillary has to be curved and it is the difference of pressure across this curved surface that has to be balanced either by the rise of water (when R is negative) or by the depression of mercury (when R is positive). Young obtained a similar expression for the internal pressure, but without the explicit link between Laplace's integrals, K and H, and the force f. Both used their results to explain all the classical phenomena of capillarity. Young called his equivalent of Laplace's ½H the surface tension, but Laplace himself never used this term. Laplace observed that K is certainly much larger than H/R, but that it is essentially unobservable since it appears in the expression for the internal pressure or stress within the liquid whatever the shape of the surface. The natural interpretation that he and Berthollet drew was that K was responsible for the cohesion of solids and liquids, and, by extension, for chemical attractions also, and that H was responsible for the weaker capillary forces. But his result was something of a pyrrhic victory in his struggle to determine the forces themselves. Knowing the value of an integral tells one nothing quantitative about the value of the integrand unless one also knows its range. He had solved the mystery of capillarity but he had not determined the forces.

There was however one interesting piece of information implicit in the integrals K and H. Their ratio, H/K, is a length which is a measure of the distance over which $f(r)$ is non-zero. If, therefore, one could make an estimate of K then the easily measured value of the surface tension, ½H, gives one a measure of the range of the forces. Since Laplace was not able to make a realistic estimate of K he could not exploit this route, but Young was more willing to attempt it. He did not approve of Laplace's hiding "the plainest truths of mechanics in the intricacies of algebraic formulas", and so arrived at this measure of the range by an argument that is less transparent than that above but essentially equivalent to it. The surface tension of water, $\frac{1}{2}H$, is 7.5×10^{-2} N m^{-1} and for water he estimated a quantity equivalent to K to be 23 kbar or 2.3×10^{9} N m^{-2} (all in modern units), by a route that is far from clear but which seems to be derived from Canton's measurement of the compressibility of water. (There is no exact equivalent to Laplace's K in modern theory. The name 'internal pressure' is now given to the rate of change of internal energy with volume, $(\partial U/\partial V)_T$, which is about 4 kbar for most simple liquids.) Young's estimate of the range of the forces of attraction was therefore $(15/2.3) \times 10^{-11}$ m, or about 0.7 Å, whereas a modern value would be closer to 5 Å. The high value of his equivalent of Laplace's K has made his result rather too small but nevertheless this is the first estimate of the range of the attractive forces that we can now recognise as being carried out by a method of approximation that is based on proper physical principles. Unfortunately Young published it anonymously in 1816 in a supplement to *Encyclopaedia Britannica* where it lay unnoticed until Lord Rayleigh discovered it in 1890, and so it brought Young no credit, nor had it any discernible effect on the progress of the field.[23]

Beyond Laplace

In the first twenty years of the 19th century Laplace dominated French physics and French physics dominated Europe. But soon the falsity of some of the assumptions made by him and his followers became apparent, and his influence declined rapidly. He had been committed to a corpuscular theory of both matter and of light. His particles of light moved at a known speed but his particles of matter were supposed to be essentially static and kept themselves apart by a balance between the attractive forces of unknown form and origin that he had exploited to explain capillarity and the repulsive force of the heat or caloric that he supposed them to carry. These views became increasingly untenable as the century progressed. The dominant mode of thought, particularly in Britain, moved away from corpuscular theories and towards field theories in which an all-pervasive aether carried the waves of heat and light and the electric and magnetic forces. The new science of thermodynamics seemed also to be quite free of any dependence on a corpuscular hypothesis. If some chemists believed in atoms, these entities seemed to have no necessary connection with the particles of the passing era of physics.

There was, however, one specialised field of mathematical physics where the faults of Laplace's theories were of little consequence; this was the field of the elasticity of solids, since, as we now know, the particles here are indeed almost static, as he supposed, and changes of temperature are of little consequence. Here both corpuscular and continuum theories grew up together. Some physicists, Cauchy for example, worked quite happily in both modes of thought. The field was, however, almost a detached area of physics that was rarely properly integrated into other parts of the subject.

Theoretical arguments showed that if crystals were indeed held together by forces between particles, then a solid of given symmetry needed fewer independent coefficients of elasticity to describe its resistance to stresses than if its cohesion were not restricted to such an origin. Thus, for example, a cubic crystal had in general three independent elastic constants, usually denoted c_{11}, c_{12} and c_{44}, but if it was held together by forces between pairs of particles then $c_{12} = c_{44}$, and so only two were independent. Such equations were called the 'Cauchy relations'. Here, then, was a possible test of whether such forces existed. From about 1840 to about 1910 the argument raged between what Karl Pearson called the rari-constant camp, who believed in inter-particle forces, hence also in the Cauchy relations, and so favoured the smaller number of elastic constants, and the multi-constant camp who insisted that the full number of constants was needed.[24] It became one of the fields of argument between those who believed in the real existence of atoms and molecules (and so of real forces between them), and those who regarded atoms only as convenient devices for rationalising some branches of physics and chemistry.

The experimental evidence was divided; thus for the cubic crystal of sodium chloride Voigt[25] found in 1888 that $c_{12}/c_{44} = 1.02$ (*i.e.*, essentially unity, as the rari-constant camp maintained), but for the cubic crystal of calcium fluoride he found a ratio of 1.32. By the end of the 19th century there was ample evidence that some materials satisfied the Cauchy relations but that most, including all the metals, did not. The engineers rightly concluded that the rari-constant hypothesis was of little use to them and briskly dismissed it as 'an error' or even as 'absurd'.[26] The matter was not resolved until 1915 when Max Born[27] showed theoretically that the reduction of the number of constants held good only if there were forces acting between pairs of particles and *also* if each particle was at a centre of symmetry of the whole crystal lattice. The argument rumbled on until the middle of the 20th century but most recognised that from the 1920s onwards we now knew how to use safely the elastic constants of some, but not all, solids to obtain useful information about intermolecular forces.

Kinetic Theory

The field theories of physics of the 19th century reached a climax with the publication in 1873 of James Clerk Maxwell's Treatise on electricity and magnetism, which, in a review in Nature, his friend Peter Tait said 'upset completely the notion of action at a distance'.[28] But at this very time evidence for forces between particles acting in just this way was accumulating rapidly from the new kinetic theories of matter.

The idea that a gas was composed of particles in rapid motion and that its pressure was a consequence of their bombardment of the walls of the containing vessel went back to the 18th century but had been revived early in the 19th by men such as Herapath, Waterston, König, and the young Joule, who were all outside the mainstream of physics. The hypothesis came into the mainstream only in 1857 when it was taken up first by Rudolf Clausius and then, paradoxically, by Maxwell himself. Arguments in its favour soon multiplied; a strong early one was Maxwell's prediction, and then experimental verification, of the unexpected fact that, over a wide range of pressure, the viscosity of a gas is independent of its density. The magnitude of the viscosity and its change with temperature gave indirect but firm evidence that the molecules had not only a size in the range of 2–5 Å, but also that they exerted forces upon each other. These forces could not be determined explicitly until the kinetic theory of gases was completed after 1916 but some results were obtained.[29] Thus Maxwell showed that if the molecules repelled each other with a force that varied inversely with the 5th power of their separation then the viscosity increased linearly with temperature. These results for the viscosity were complemented by measurements of the departure of real gases from the ideal laws of Boyle and Charles. Such departures could most simply be interpreted, qualitatively but not then quantitatively, as the effect of intermolecular forces. The conflict of this evidence with that in favour of field theories, or for a continuum treatment of the elasticity of solids, was rarely faced. The pragmatic physicists of the day were generally glad to be able to leave problems of this kind to be sorted out later.

The amalgamation of the new kinetic theory of gases with Laplace's handling of the attractive forces between the molecules in a fluid was accomplished in a thesis[30] by a hitherto unknown Dutch schoolmaster, J. D. van der Waals. This was published in 1873, the same year that Maxwell published his *Treatise*. Van der Waals's work, *On the continuity of the gaseous and liquid states*, became best known for the famous equation of state of gases and liquids that formed the conclusion of Chapter 7;

$$(p + a/V^2)(V - b) = RT, \qquad (3)$$

where p, V and T are pressure, molar volume and temperature, where R is the universal gas constant, and where a and b are parameters characteristic of each substance. It was equally important, however, that he brought the study of intermolecular forces, apparently acting at a distance, back into the mainstream of physics. He was no Boscovichian; he treated molecules as hard bodies between which there were attractive forces of short range. He estimated the size of these hard bodies from the parameter b, the excluded volume of his equation, and Laplace's integral K from the parameter a, where $K = a/V^2$. His equation is therefore an expression of the fact that if we take the total stress in a fluid, that is, the sum of the hydrostatic pressure, p, and the internal pressure,

a/V^2, and multiply this sum by the effective molar volume open to the molecules, $(V - b)$, then we obtain the familiar expression, RT.

Van der Waals then repeated Young's argument, of which he knew nothing, to obtain the range of the attractive forces from the ratio H/K. He found that the range, 3–5 Å, was about the same as the apparent diameter of his hard bodies, from which he deduced that molecules attracted each other only when they were exceedingly close together. In these deductions and in his estimate of molecular size, he was to be proved to be essentially correct, and, even in his day, his work convinced many of his contemporaries that such an interpretation was sound. His equation of state is only a crude approximation to that of real gases and liquids but the physical principles on which it is based are generally sound, and so his deduction from it of the size and forces of molecules are close to our modern estimates.

Where did this leave those who could not accept forces acting through empty space, even for a short distance? Most shrugged their shoulders and again left it to the future to solve the problem. Maxwell, with a foot in both camps, and a strong interest in philosophical problems, was one who was not happy to see the problem ignored and in a detailed argument in a Friday Evening Discourse at the Royal Institution in 1873 he attempted to resolve the difficulty, but without coming to a firm conclusion: "If we are ever to discover the laws of nature, we must do so by obtaining the most accurate acquaintance with the facts of nature, and not by dressing up in philosophical language the loose opinions of men who had no knowledge of the facts".[31] The problem of action at a distance was not to be solved until it was known exactly how the forces arose.

The Origin of the Forces

By the end of the 19th century those who could not accept the reality of atoms and molecules, and of the forces between them, were becoming a dwindling minority. But those who did believe in them still could not say what gave rise to the forces. With the discovery of the electronic structure of atoms, electric explanations became fashionable in the opening years of the 20th century. Thus a molecule such as that of water was known to have an electric dipole; the hydrogen atoms were believed, correctly, to carry small net positive charges that were balanced by a negative charge associated with the oxygen atom. More complicated electric structures were envisaged for other molecules, and it was not difficult to show that the result of such an array of molecular charges is a weak attractive intermolecular force. But quantitatively the sums did not add up; something important was missing. Moreover the inert gases had recently been discovered and there was good reason to think that their molecules were spherically symmetrical atoms. Such a neutral spherical atom could not have a field around it, according to the accepted laws of classical electrostatics, but it was known that argon condensed to liquid and froze to a solid quite as readily as nitrogen, oxygen or methane. So some more general source for the force had to be found.

The problem was insoluble within the laws of classical electricity and magnetism. The evidence that something was seriously wrong with these laws when applied to atomic systems accumulated from the late 19th century onwards. *Ad hoc* attempts by Planck, Einstein, Bohr and others to patch up the classical laws, generally known as the 'old quantum theory', led to improvements in the interpretation of atomic spectra, and in some other branches of physics, but they still told us nothing about the intermolecular forces. The solution came only in the years between 1927 and 1930 with the development of the 'new' quantum theory that we now recognise as the proper tool for handling the physics of atomic systems.

At the suggestion of the Dutch physicist, Peter Debye, who was visiting Columbia University in New York, a Chinese graduate student there undertook in 1927 the quantal calculation of the energy of a pair of hydrogen atoms that were close together and whose electrons were restricted to move in the same plane. Shou Chin Wang[32] found that each atom was circularly symmetrical when they were far apart, and so exerted no forces on each other, but when they were close together there was indeed a small attractive force that arose from the coupling of the motions of the electrons on each atom around the positive nuclei. The force fell off rapidly with increasing separation, as the inverse 7th power of the distance; it was of the right order of magnitude to account for what was known of the strengths of intermolecular forces; and its short range confirmed all the evidence that had accumulated from the time of Hauksbee. The force is essentially quantal in origin; there is no classical analogue to the coupling of the electronic motions.

Wang's crude model and far from exact treatment was soon superseded by the more convincing calculations of Fritz London and Robert Eisenschitz on a pair of properly three-dimensional hydrogen atoms. It is London's name that we now often associate with these weak attractive forces of short range. They are, however, also called dispersion forces, from the electrons' role in causing the dispersion of light, and sometimes van der Waals forces, although that name is usually extended to cover both these forces and the other miscellaneous electrostatic forces that also contribute to cohesion. All these forces are weak in comparison with those that lead to chemical bonding, which the new quantum theory was equally successfully in describing, but they are, nevertheless, responsible for many of the most tangible properties of matter.[33]

The quantal calculations showed the origin of the forces and gave an estimate of the quantitative strength of the sole or most important component of the intermolecular attraction in simple molecular systems such as argon, nitrogen or chlorine but, because of the difficulty of the calculations, they did not solve the problem of measuring the strength of the forces in such systems, nor of determining the relative importance of these and the classical electrostatic forces in more complicated systems. That task was looked at in the 1930s but undertaken seriously only after the Second World War. It has proved to be surprisingly difficult; it was only in 1971 that most of those working in the field were convinced that one of the simplest cases, the experimental and theoretical determination of the force between a pair of argon atoms, had been solved.[34] Progress with more complicated systems has been slow. There has recently been great progress in the important case of water by a detailed analysis of the infrared spectrum of the water dimer[35] and this technique can be extended with difficulty to other more complicated systems. Quantitative progress is not going to be easy but the principles are clear; we have reached the position envisaged by Boltzmann in 1895, when he looked forward to the time when 'the difficulty of enumerating all the material points of the universe, and of determining the law of mutual force for each pair, would only be a quantitative one; nature would be a difficult problem, but not a mystery for the human mind'.[36]

Action at a Distance

London's mechanism is an electromagnetic one and so it is implicit in his theory that the attractive force is not transmitted instantaneously from one molecule to another but only at the speed of light. The separation of molecules in a typical liquid or solid is of the order of $3–6 \times 10^{-10}$ m, and since the speed of light is 3×10^{8} m s^{-1}, the time for the transmission of the force from one molecule to its neighbour is only about $1–2 \times 10^{-18}$ s, or little more than an attosecond, a duration still outside the range of modern chemistry and physics. For most purposes this can be taken to be an instantaneous propagation and little was said in the early days about this duration or about the mechanism of the transmission of the force. It similarly came to be accepted that, as is required by the general theory of relativity, the gravitational force was propagated at the speed of light by hypothetical massless particles now christened 'gravitons'. But although the relevant distances and times are large for the Sun and planets, there is, I believe, still no direct experimental evidence for the finite speed of the gravitational force.[37] Whether gravitons are, in Newton's phrase 'material or immaterial' agents is a question that I shall not enter into.

During the 1930s a series of experiments on colloids by a group in the Phillips Laboratories at Eindhoven led to the conclusion that the London forces between the many molecules involved played a major role in determining the stability of suspensions of colloidal particles. J. Th. G. Overbeek found, however, that the force between a pair of colloidal particles was much smaller than the force that he had calculated by adding all the contributions from each pair of molecules, one in the first particle and one in the second. He thought that this weakening might be due to the non-zero time taken for the force to be transmitted from one particle to the other across a distance that was large on a molecular scale. He put this point to his colleagues H. B. G. Casimir and D. Polder who carried out the difficult quantal calculation, which required fourth-order perturbation theory, and found that there was indeed a 'retardation' of the force. What was a dependence on distance proportional to the inverse 7th power at short distances became a weaker force with a dependence on the inverse 8th power at macroscopic distances.[38] The characteristic length at which this retardation sets in for a pair of molecules of ionisation potential I is $hc/2\pi I$, where h is Planck's constant and c is the speed of light. This distance is 125 Å for argon and a similar length for other molecules. It is, therefore, so far beyond the effective range of the London force between a pair of atoms, about 5 Å, that is irrelevant. It is, however, important for a pair of colloidal particles of a diameter and separation of about 1 µm, or 10^4 Å. If the speed of light were infinite then there would be no retardation since the characteristic length at which it sets in would also be infinite.

This indirect demonstration of the existence of the retardation of the forces by its effect on colloid stability and by theoretical calculations was satisfactory as far as it went but a direct measurement was more difficult. There were attempts at measuring the attraction between smooth glass spheres by Overbeek in the Netherlands and by B. V. Deryagin in Moscow, but it is impossible to make the spheres sufficiently smooth for them to be brought close enough together to measure the force with adequate accuracy. The breakthrough came in 1968 when David Tabor in Cambridge exploited the fact that he could cleave mica sheets to yield surfaces that were atomically smooth over distances of several millimetres. By measuring the force of attraction between two sheets curved at right angles to each other he was able to show that the London force was 'normal' to a separation of about 100 Å and was fully 'retarded' beyond 200 Å, in accord with the theoretical calculations.[39] With this demonstration of the consequence of the electromagnetic mechanism and of the finite speed of the transmission of intermolecular forces the problem of action at a distance has at last been laid to rest, at least in this context.

Conclusion

The problem of the cohesion of matter was, we now know, formulated correctly by Newton and his followers at the start of the 18th century but it ran into opposition because neither he nor they could say how the forces acted and because of metaphysical objections on the Continent to action at a distance. Little was accomplished throughout that century, mainly, I think, because there were the more exciting fields of electricity, magnetism, heat and light to explore, and because the natural philosophers of the day were either not convinced of the correctness of Newton's assumptions (Euler, for example[40]) or because their skills, mathematical or physical, were not adequate to tackle the problems (Clairaut[41]). There was a dramatic advance at the start of the 19th century when Young and, more particularly, Laplace combined a belief in particles and the forces between them with a mathematical competence sufficient to tackle the crucial problem of capillarity. Young, in particular, was no better a mathematician than many of his 18th century predecessors and there is no compelling rational reason for the timing of their sudden solution to the problem. In Young's case his interest may have been a consequence of his reverence for Newton and his works, and Laplace's interest was aroused by his friend Claude-Louis Berthollet who, like Freind a hundred years earlier, wished to interpret chemistry in terms of Newtonian attractions. A second cause of Laplace's interest was his need to interpret the refraction of the light as an attraction of its particles by the particles of matter. But the timing of their common solution, over a period of twelve months, to a well-known problem that had been outstanding for over a century is probably not capable of any entirely rational explanation.

The basic physical notions of 1800 – the caloric theory of heat, a corpuscular theory of light, and the static picture of the particles of matter – were, however, still inadequate to support further progress. The field again dropped out from the mainstream of physics under competition from the thriving work on electricity and magnetism, on light, and on heat, in the form of the new science of thermodynamics. It was only with the general acceptance of a kinetic theory of molecular matter in the second half of the 19th century that the subject came back into fashion, only to stumble again when the current theories of electricity proved unable to produce a convincing explanation of the origin of the forces. It was not until the new quantum theory of the late 1920s that this block was finally overcome in principle, and it was another 40 years before theory and experiment could combine to solve quantitatively even so simple a case as that of the force between two argon atoms. The science of statistical mechanics, the more wide-ranging successor to the kinetic theory of gases, then led to the successful use of these forces to tackle quantitatively the calculation of the properties of simple forms of matter in all three states: gas, liquid and solid. These successes still leave vast fields to be tackled such as the application of the theories to explain quantitatively the interaction of molecules and ions in biological systems and in inorganic materials of practical importance, but, as Boltzmann put it, the problems before us are clear and the field is no longer a mystery.

I believe that the history of cohesion provides, however, an instructive example of the irregular and adventitious progress of what Thomas Kuhn called 'normal science'. Progress stops and re-starts, sometimes because of new discoveries, either theoretical or experimental, but also because of competition from more rapidly moving fields, or even for reasons that are often subsumed under the word 'contingency' but which, in truth, cannot be described as more than fashion. "Everything has its fashions, even philosophy has its own", wrote the French polymath Réaumur[42] in 1749. The same is true today but, because of the many scientists now active throughout the world, a field is never wholly neglected. The study of intermolecular forces is today probably not as fashionable as it was thirty years ago but will doubtless return to favour in its own good time, probably through the increasing interest in the interaction of biological molecules and the increasing use apparatus that allows the measurement of forces on an atomic scale.

NOTES

1. E. J. Aiton, *The vortex theory of planetary motions*, Macdonald, London, 1972.
2. [I. Newton], *Four letters from Sir Isaac Newton to Doctor Bentley containing some arguments in proof of a deity*, London, 1756; reprinted in *The correspondence of Isaac Newton*, 7 volumes, University Press, Cambridge, 1959–1977, vol. 3, pp. 253–256. A fuller exposition of the work of Newton and his followers, and of later developments down to the end of the 20th century, may be found in J. S. Rowlinson, *Cohesion: a scientific history of intermolecular forces*, University Press, Cambridge, 2002.
3. J. Locke, *An essay concerning human understanding*, 4th edn, London, 1700, Book II, chap. 8, §11; M. B. Hesse, *Forces and fields: the concept of action at a distance in the history of physics*, Nelson, London, 1961, p. 49.
4. Traces of these ideas can be found in Newton's *Principia*, but they did not emerge clearly until the Latin edition of his Opticks in 1706, and are expressed most fully in Query 31 of the 4th edition of 1730, reprinted by Dover, New York, 1979, pp. 375ff.
5. I. Newton, unpublished notes, transcribed by I. Cohen, '*Hypotheses in Newton's philosophy*', *Physis, 8* (1966) 163–184.
6. J. Freind, *Chymical lectures: In which almost all the operations of chymistry are reduced to their true principles and the laws of Nature*, London, 1712, pp. 172–200. The Latin edition was published in 1709.
7. F. Hauksbee, *Physico-mechanical experiments on various subjects*, 2nd edn, London, 1719, pp. 194–217; J. Jurin, 'An account of some experiments shown before the Royal Society; with an enquiry into the cause of the ascent and suspension of water in capillary tubes', *Phil. Trans. Roy. Soc.* **30** (1718) No. 355, 739–747.
8. [B. Taylor] 'Extract of a letter from Dr. Brook Taylor ... Giving an account of some experiments relating to magnetism', *Phil. Trans. Roy. Soc.* **31** (1721) No. 368, 204–208.
9. S. Hales, *Vegetable staticks,* London, 1727; J. T. Desaguliers, 'Some thoughts and conjectures concerning the cause of elasticity', *Phil. Trans. Roy. Soc.* **41** (1739) No. 454, 175–185.
10. R. J. Boscovich, A theory of natural philosophy, Open Court, Chicago, 1922, § 266, p. 205, a translation of the 2nd edn of *Theoria philosophiae naturalis,* Venice, 1763.
11. J. Canton, 'Experiments to prove that water is not incompressible', *Phil. Trans. Roy. Soc.* **52** (1762) 640–643.
12. J. Priestley, *The history and present state of discoveries relating to vision, light and colours,* 2 volumes, London, 1772, v. 1, pp. 390–393, and in a fuller account five years later, in a chapter entitled 'Of impenetrability, as ascribed to matter' in *Disquisitions relating to matter and spirit,* London, 1777, pp. 11–23. See also, L. P. Williams, 'Boscovich and British chemists', pp. 153–167, and R. E. Schofield, 'Boscovich and Priestley's theory of matter', pp. 168–172, in L. L. Whyte, ed., *Roger Joseph Boscovich*, Allen & Unwin, London, 1961, and R. E. Schofield, 'Joseph Priestley and the physicalist tradition in British

chemistry', pp. 92–117, in E. N. Hiebert, A. J. Ihde and R. E. Schofield, ed., *Joseph Priestley: Scientist, Theologian, and Metaphysician*, Associated Universities Press, Cranbury, NJ, 1974.

13. J. Michell, *A treatise of artificial magnets ...*, Cambridge, 1750.

14. J. Rowning, *A compendious system of natural philosophy ...*, Pt. 2, London, 1738, pp. 5–6, 65–72; [J. Priestley], *Memoirs of Dr. Joseph Priestley, to the year 1795, written by himself: with a continuation, to the time of his decease, by his son, Joseph Priestley, ...*, London, 2 volumes,1806, 1807, vol. 1, pp. 183–184.

15. R. E. Schofield, 'Joseph Priestley: Natural Philosopher', *Ambix* **14** (1967) 1–15; J. G. McEvoy, 'Joseph Priestley, Natural Philosopher: some comments on Professor Schofield's views', *Ambix* **15** (1968) 115–123; J. G. McEvoy, 'Causes and laws, powers and principles: The metaphysical foundations of Priestley's concept of phlogiston', pp. 55–71, in R. G. W. Anderson and C. Lawrence, ed., *Science, Medicine and Dissent: Joseph Priestley* (1733–1804), Wellcome Trust, London, 1987.

16. J. Priestley, *Memoirs*, ref. 14, vol. 1, pp. 78–79: M. Barfoot, 'Priestley, Reid's circle and the Third Organon of Human Reasoning', pp. 81–89 in Anderson and Lawrence, ref. 15.

17. Lord Kelvin, 'Plan of an atom to be capable of storing an electrion with enormous energy for radio-activity', *Phil. Mag.* **10** (1905) 695–698.

18. J. Leslie, 'On capillary action', *Phil. Mag.* **14** (1802) 193–205.

19. G. Monge, 'Mémoire sur quelques effets d'attraction ou de répulsion apparente entre les molécules de matière', *Mém. Acad. Roy. Sci.* (1787) 506–529, published in 1789.

20. N. Desmarest, in his French edition of F. Hauksbee, *Expériences physico-méchaniques sur différens sujets*, Paris, 1754, vol. 2, p. 137.

21. T. Young, 'An essay on the cohesion of fluids', *Phil. Trans. Roy. Soc.* **95** (1805) 65–87.

22. P. -S. Laplace, 'Sur l'action capillaire', a separately paged supplement of 1806 to vol. 4 of *Traité de mécanique céleste*, Paris, 1805. This was followed by a second 'Supplément à la théorie de l'action capillaire' of 1807, usually bound into the same volume.

23. [T. Young], 'Cohesion', *Supplement to the fourth, fifth, and sixth editions of Encyclopaedia Britannica*, London, 1816, vol. 3, pp. 211–222; Lord Rayleigh, 'On the theory of surface forces', *Phil. Mag.* **30** (1890) 285–298, 456–475.

24. I. Todhunter and K. Pearson, *A history of the theory of elasticity*, University Press, Cambridge, 1886–1893, vol. 2, pt. 2.

25. W. Voigt, 'Bestimmung der Elasticitätsconstanten von Flusspath, Pyrit, Steinsalz und Sylvin', Ann. Physik 35 (1888) 642–661.

26. L. L. Bucciarelli and N. Dworsky, *Sophie Germain: an essay in the history of the theory of elasticity*, Reidel, Dordrecht, 1980, pp. 66, 71.

27. M. Born, *Dynamik der Kristallgitter*, Teubner, Leipzig, 1915; M. Born and K. Huang, *Dynamical theory of crystal lattices*, Clarendon Press, Oxford, 1954.

28. J. C. Maxwell, *A treatise on electricity and magnetism*, Clarendon Press, Oxford,

1873; [P. G. Tait], 'Clerk-Maxwell's Electricity and Magnetism', *Nature* **7** (1873) 478–480.

29. S. Chapman and T. G. Cowling, *The mathematical theory of non-uniform gases*, University Press, Cambridge, 1939.

30. J. D. van der Waals, *On the continuity of the gaseous and liquid states* [1873], English translation in *Studies in Statistical Mechanics*, North-Holland, Amsterdam, vol. 14, 1988.

31. J. C. Maxwell, 'On action at a distance', *Proc. Roy. Inst.* **7** (1873) 44–54.

32. S. C. Wang, 'Die gegenseitige Einwirkung zweier Wasserstoffatome', *Phys. Zeit.* **28** (1927) 663–666.

33. F. London, 'Über einiger Eigenschaften und Anwendungen der Molekülarkräfte', *Zeit. phys. Chem.* **B11** (1930) 222–251; R. Eisenschitz and F. London, 'Über das Verhältnis der van der Waalsschen Kräfte zu den homöopolaren Bindungskräften', *Zeit. f. Phys.* **60** (1930) 491–527. London first gave a simplified explanation of his calculation in English in 'The general theory of molecular forces', *Trans. Faraday Soc.* **33** (1937) 8–26, reprinted, with a commentary by A. D. Buckingham, in *100 Years of Physical Chemistry*, Roy. Soc. Chem., Cambridge, 2003, pp. 1–21.

34. G. C. Maitland, M. Rigby, E. B. Smith and W. A. Wakeham, *Intermolecular forces: their origin and determination*, Clarendon Press, Oxford, 1981.

35. R. S. Fellers, C. Leforestier, L. B. Braly, M. G. Brown and R. J. Saykally, 'Spectroscopic determination of the water pair potential', *Science* **284** (1999) 945–948.

36. L. Boltzmann, 'On certain questions of the theory of gases', *Nature*, **51** (1895) 413–415.

37. D. Keeports, 'Why c for gravitational waves?', *Amer. Jour. Phys.* **64** (1996) 1097, and five answers to this question, ibid. **65** (1997) 589–592.

38. H. B. G. Casimir and D. Polder, 'Influence of retardation on the London-van der Waals forces', *Nature* **158** (1946) 787–788; *Phys. Rev.* **73** (1948) 360–372; E. J. W. Verwey, and J. Th. G. Overbeek, *Theory of the stability of lyophobic colloids*, North-Holland, Amsterdam, 1948.

39. D. Tabor and R. H. Winterton, *'Surface forces: Direct measurement of normal and retarded van der Waals forces'*, Nature *219* (1968) *1120–1121; 'The direct measurement of normal and retarded van der Waals forces'*, Proc. Roy. Soc. *A 312 (1969) 435–450; J. N. Israelachvili and D. Tabor, 'The measurement of van der Waals dispersion forces in the range 1.5 to 130 nm'*, ibid. *331 (1972–1973) 19–38.*

40. L. Euler, *Lettres à une princesse d'Allemagne sur divers sujets de physique et de philosophie*, St Petersburg, 1768, Letter No. 79, vol. 1, pp. 312–314.

41. A. Clairaut, *Théorie de la figure de la Terre*, Paris, 1743, pp. 105–128. This chapter records an unsuccessful attempt at the calculation that Laplace made successfully in 1806.

42. R. A. F. de Réaumur, *Art de faire éclore et d'élever en toute saison des oiseaux domestiques de toutes espèces*, Paris, 1749, vol. 2, p. 328.

SECTION 10

BOC Symposium: Physical Science

JOSEPH PRIESTLEY AND THE 18TH CENTURY HISTORY OF ELECTRICITY

Peter Excell
University of Bradford

For Joseph Priestley (1733–1804) and his contemporaries, electricity meant 'static electricity' although the magnetisation of iron needles had allegedly been shown in the 18th century. Experiments on the effect of static electricity on frog legs were performed by Luigi Galvani (1737–98) in 1791 and the pile of Alessandro Volta (1745–1827) was invented in 1800, but 'Galvanism' was still regarded as a separate phenomenon in an 1823 textbook. Michael Faraday (1791–1867) propounded his Law of Electromagnetism in 1831. Optics was not linked with electricity and magnetism until James Clerk Maxwell (1831–79) in the 1870s.

Peter Excell

Early days

The following summarises key events before Priestley's birth:

- 1662 – Boyle's Law (compression of gases); experimental science; study of gases
- 1663 – Von Guericke's Air Pump; 'Magdeburg hemispheres' experiment shows reality of air as a gas; also did some electrical work
- ca. 1666 – Newton's Laws (motion and gravity); the rise of modern science; the concept of force acting at a distance

There was great difficulty in getting consistent results in electrical (*i.e.*, electrostatic) experiments due to widely varying conductivities of apparently identical materials, Magnetism however was much more consistent in its behaviour and gravity even more so. There was a failure to appreciate the significance of moisture/humidity and there were difficulties in generating charge due to surface subtleties. On a positive point though, electricity was free of dogma (mainly religious) that plagued astronomy (*c.f.*, Giordano Bruno, Galilei Galileo, etc).

17th Century

At the beginning of this period, there was no ethos of experimentation but the subsequent invention of the air pump provided a key tool for experiments and demonstrations such as those of Otto von Guericke (1602–86). Jean Desaguliers (British-Huguenot, 1683–1744) was another leading experimenter of the time. The existence of electric fishes was known and these were linked to electricity. Francesco Lana (1631–87), a Jesuit, noted in 1670 that "a magnet draws iron and not straw: amber [vice versa]"; field analogies were being drawn from this. Corpuscular theories ('electrical effluvia') were presented by René Descartes (1596–1650) to explain attraction as complex screw actions in the interstices of a particle matrix. In 1706, Isaac Newton (1642–1727) proposed that electricity might figure in the production of light – hence that electrical 'effluvia' might be a fundamental force of Physics. He also considered that air consisted of particles repelling each other – a hint of electrical forces? – but stated that he had no idea how it happened. The concept of 'action at a distance' was an overall philosophical problem, tending to occultism. Most disliked it, often postulating invisible streams of particles (effluvia) to circumvent the occult.

18th Century

Francis Hauksbee (1666–1713) was nominated by Newton to be a demonstrator and his personal assistant at Royal Society – both joined the Society in 1703. At Newton's wish, the Society was revived by the performance of experiments. It is possible that Hauksbee invented the so-called 'globe electrical machine'. Demonstrations of glows at low pressure were presented at Royal Society meetings in 1705 by rubbing the outside of an evacuated globe; at higher pressure sparks were produced – note that this essentially links gases, electricity and light. However, observers were perplexed by the behaviour of sensing threads inside the globe when a finger approached – not explicable by effluvia theory. Furthermore, a glow was observed in a globe surrounded by another being rubbed and this was explained by effluvium passing through glass, but this effect was screened by muslin and there was no obvious reason for this. Also it was noted that the field (effluvium) from sealing wax would pass through glass so that the idea that effluvia from one material would pass only through like materials could not be right.

Christiaan Huygens (1629–95) proposed the existence of a 'luminiferous ether' as a necessity for his wave theory of light thereby hinting at a connection of electricity with light; vortices in the ether were then used to explain forces. Repulsive forces were certainly less obvious in electrostatics than in magnetism. The existence of oscillating discharging objects caused a major problem for theories.

Stephen Gray (1666–1736), an acquaintance of John Flamsteed the first English Astronomer Royal, worked on induced electrification and circuits; his observations undermined effluvia theory. Gray may have been the discoverer of insulation.

Charles Dufay (1698–1739) was credited with the discovery of two different types of electricity (from a comparison of rubbing glass and amber, etc); these were regarded as elements or possibly as species of 'fire'. In the 1730s, Dufay recognised electrostatic repulsion as qualitatively different from gravitation.

Roger Boscovich (1711–87) believed that repulsive forces applied over short distances, then attractive forces at greater distances, then repulsive forces at greater distances still, etc.

In summary, by early 18th Century, the existence of two types of electricity, and of conductors and insulators, were understood. There was a consequent need for a 'theory of electricity'. Based on the general idea of 'effluvia', there were many theories in 1740s involving particles shot out by bodies. These theories were subsequently undermined by the invention of the Leyden Jar (a form of condenser) by Ewald von Kleist (a former student at the University of Leyden in Holland, 1700–48) in 1745, leading to concepts of capacity, charge and tension. Von Kleist recognised the condenser as a portable source of electricity (although he did not understand how it worked).

- 1733: Joseph Priestley born at Fieldhead (Owler Lane), a hamlet near Birstall[1]

Priestley birthplace at Fieldhead

In 1746, the Leyden jar was used by Pieter van Musschenbroek (1692–1761) to produce large ("near fatal") shocks. Priestley criticised van Musschenbroek for fearing to repeat the experiment; *c.f.*, Georg Bose who said he wished he might die of the electric spark, to do something notable (a joke misinterpreted by Priestley as being 'serious'!). Around this period, no one understood how the Leyden jar worked; it broke the 'rule of Dufay' in that the more insulation the better the Jar worked as an electricity store.

This period essentially marked a watershed in the understanding of stored electricity. Subsequent to this, there were several significant developments:

- Stronger sparks from Leyden jars increased their linkage to lightning – similar because both conducted by metals.
- The parallel plate condenser was invented; surprisingly, it was found that thinner glass was better than thicker.
- Varying the distance from the base of a Leyden jar to ground varied its "power".
- A widespread early view of the Jar was that the glass acts as a "pump" of electric atmospheres.

By 1745, there was talk of 'male and female electricities', but the speed of various effects was too fast to measure (William Watson 1748).

- ca. 1750: Priestley starts educating himself in the sciences
- 1752 (age 19): Priestley enters Daventry Academy for dissenting ministers

Benjamin Franklin (1706–90) initially considered electricity to be of two types, named positive and negative, and thought of these as atmospheres or fires, but subsequently he moved to the view that there was only one type, with a deficiency of itself as its opposite. The Jar has to balance quantities of "fire" on opposite sides of the glass. Franklin was leader of the so-called Philadelphia Group, where a colleague showed that the Jar could be electrified in reverse, a new observation. It was found that the charge appeared to reside in the glass and not the coating. Repulsion between negatively charged bodies was a problem for Franklin as the "absence of electricity" should not do anything. He thought glass to be impermeable to electric atmospheres because the pores in it shrink during cooling, but only in the centre of the glass! This caused a problem when it was demonstrated that half could be ground away and the Jar still worked.

A Digression on Lightning

In 1749, Jean-Antoine Nollet (1700–70) and Franklin listed similarities between 'electricity' and lightning. Franklin made plans to 'draw' lightning, to test its properties; this was actually done by the French in 1752.

The experiment at Marly according to a 19th century artist

Wilson's demonstration of the superiority of knobs

There was much superstition about lightning, believing that ringing bells or firing cannons could disperse thunder and lightning – many bell-ringers were killed! The question was raised as to how clouds became charged, negatively to Franklin's surprise. There was substantial controversy as to whether a pointed lightning conductor was better than one ending in a knob to convey lightning to earth if it happened to strike. In 1772, this controversy came to a head. When considering the lightning protection of the Purfleet Arsenal and the Royal Palace, Franklin favoured points as did the Royal Society committee set up by the King, but the Arsenal was still hit by lightning. Benjamin Wilson (English portrait painter and electrician, 1721–88) favoured knobs and had experiments funded, convincing the King that knobs were superior. The issue was embroiled in politics, especially as Franklin was a Republican. After several fatalities among experimenters, experiments on lightning were stopped and work moved to lightning protection.

In 1746, Nollet tried the medical application of the Leyden jar on paralytics. He generally opposed Franklin's theories, believing that 'fiery electrical matter' passed through glass; supporting this was the fact that glows were observed on the outside of a jar in vacuum.

There was strong rivalry between Philadelphia and Europe (especially Paris) on electricity theories.

Giambatista Beccaria (1716–81) wrote in 1753 a comprehensive treatise on artificial and natural electricity. This attempted clarification of positive and negative electricity, partly following Franklin but trying to advance his theory; it disagreed with Nollet's single fluid theory. Ideas on the positive and negative problem were driven by observed differences in discharges; the positive side

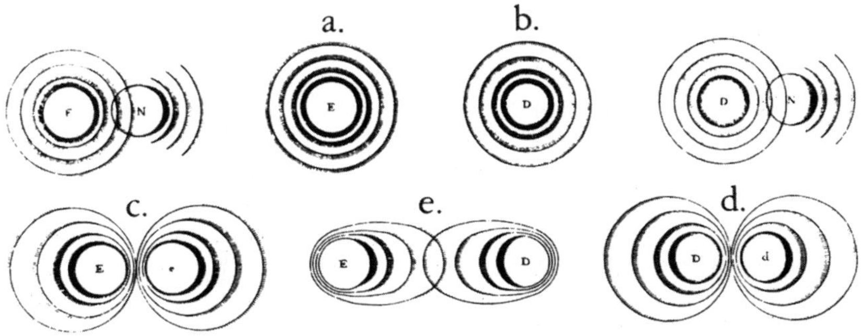

Beccaria's representation of the electric field: (a) About a positive body, (b) about a negative one, (c) between positive bodies, (d) between negative ones, (e) between unlike ones. NB These are close to modern concepts of electric field distributions, apart from crossing of field lines in (e) and the top left and right diagrams

appeared more active (*c.f.*, later work by Crookes and Faraday). Franklin and Priestley thought both sides of the jar were positive. A major problem was the apparent single type of electricity associated with glass, and hence how could the glass in the jar contain both types? In 1753, John Canton (1718–72) discovered the phenomenon of electrostatic induction.

- 1755: Priestley becomes Minister in a chapel at Needham Market, Suffolk

In 1756, Johan Wilcke (1732–96) and Franz Aepinus (1724–1802) discovered the air condenser, thereby upsetting Leyden jar theories. They suggested an 'agnostic' approach to accepting repulsive electrical forces without questioning the mechanism (*c.f.*, gravity). It was found that the mineral tourmaline could attract objects rather like lodestone; Aepinus decided (1756) that the force was electrical. In 1759, he suggested the inverse square law, but very tentatively. Whilst working in St Petersburg, Aepinus then produced a major book on electricity and magnetism *Tentamen theoriae electricitatis et magnetismi*, a very mathematical work in Latin, not understood by many.

- Benjamin Franklin's mission to England 1757–62 as Emissary of Colonies, but simultaneously pursuing scientific and political ideas
- 1758: Priestley becomes a Minister at Nantwich, Cheshire, setting up a school there and acquiring 'philosophical [*i.e.*, scientific] instruments'

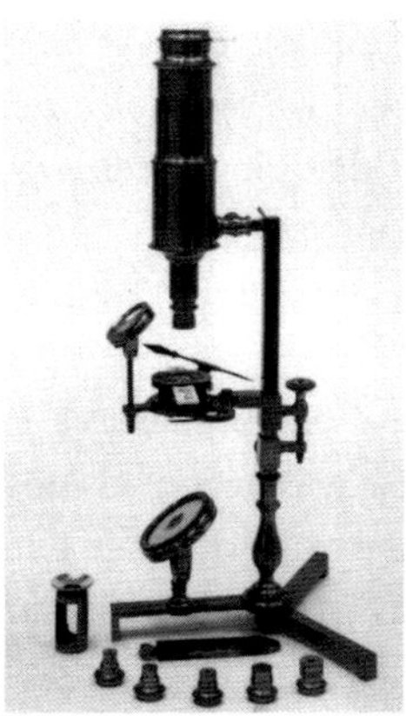

Priestley's microscope and vacuum pump at Nantwich

Nollet's elucidation of Symmer's socks. (a) When apart, the stockings swell owing to collisions between effluents from their internal surfaces, and 'attract' exterior bodies P by the actual effluent, (b) they collapse when combined because of the interference between superposed and vitreous jets. Note the short range of the jets between CD and EF

a.

b.

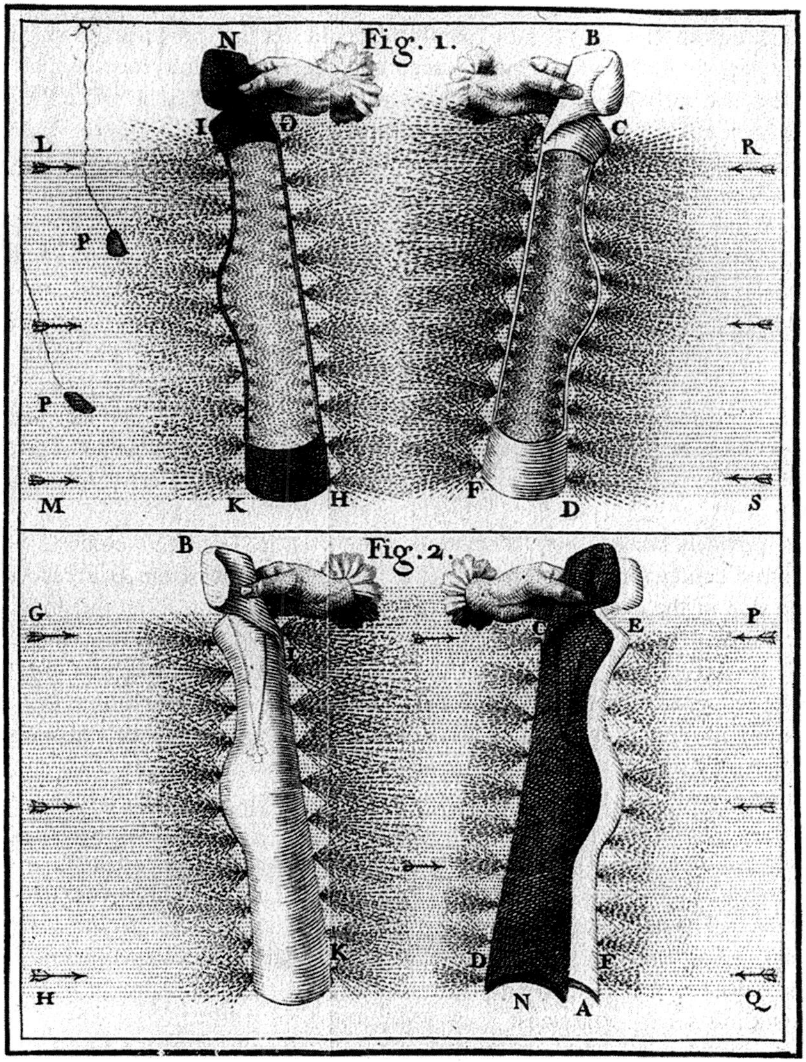

In 1758, there were the bizarre observations on 'Robert Symmer's socks'; these appeared to be oppositely charged when pulled off, thus favouring two contrary electricities. The work, published in the Philosophical Transactions of the Royal Society (!), was dismissed by Priestley since it tended to be anti-Franklin, but it was adopted by Nollet (desperate to revive his theory).

- 1761: Priestley becomes tutor in languages and literature at Warrington Academy, a nonconformist higher-education establishment. Whilst there, he makes contacts with Franklin, who encourages him to write *The History and Present State of Electricity* (HPSE), a complete compendium of electrical knowledge.
- 1766: HSPE ("History and Present State of Electricity") published in 2 volumes.[2] It contains much on sparks and electrostatic generators. The

Warrington Academy 1761

(below) Robison's equipment

(bottom) Coulomb's apparatus for measuring $f(r)$: (a) The torsion balance, (b) the electrometer stalk, (c) detail of the knob by which the wire is twisted, (d) the probe

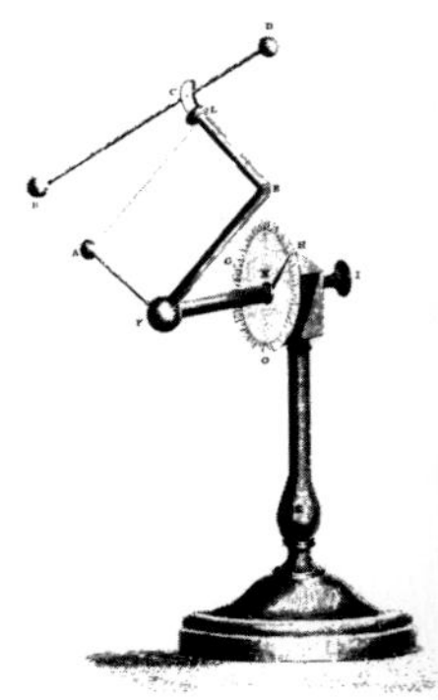

book is very qualitative in its treatment, but noteworthy in predating the concept of electric circuits. It was still influencing teaching over 100 years later, as noted for example in pages of Oliver Lodge's notebooks whilst a student in Liverpool. Sparks (discharges) have remained a source of fascination until recently, but it was the user-friendly approximations of circuit theory that really liberated electricity. On almost the last page of HPSE, Priestley proposes what is now known as Coulomb's Law of Electrostatic Force Fields, drawing on Newtonian gravitational field theory; this law is fundamental to the nature of matter.

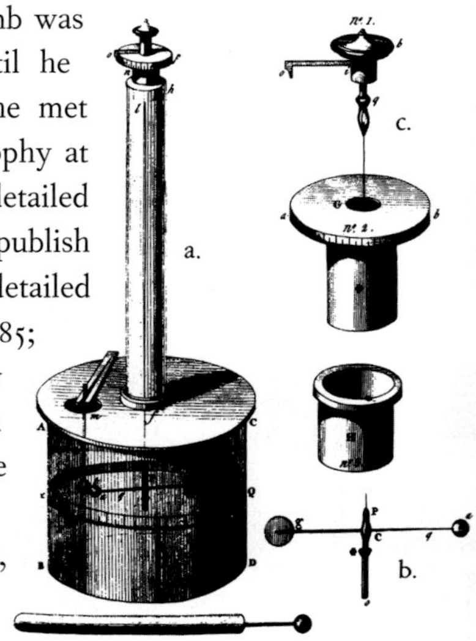

Another person who 'discovered' Coulomb's law before Coulomb was John Robison (1739–1805), who had a chequered career until he obtained an academic post at St Petersburg in 1769 where he met Aepinus. He subsequently became Professor of Natural Philosophy at Edinburgh (1774–1805). Around 1769, he performed quite detailed inverse square law experiment but, plagued by self-doubt, did not publish until 1822! Charles-Augustin de Coulomb (1736–1806) after very detailed experiments published the law customarily accredited to him in 1785; he favoured two electricities. The Law, which was only gradually accepted, was publicised in English by Robison in Encyclopaedia Britannica. In the form of the law proposed, point charges were required and this was not the obvious experimental geometry.

Electric winds appeared to be evidence of electrical fluids, although some interpreted it as air expelled by the electrical force (now considered correct).

In the second half of the 18th Century, there was considerable controversy about whether electricity was two fluids or one. Franklin favoured one, together with the concept of a "deficiency", but had a problem with negative repulsion. Robert Symmer (1707–63) and the Europeans favoured two. Priestley wrote "no fact can be shown to be absolutely inconsistent with either". This was a fair assessment, not obsessively pro-Franklin. For all theories, there was a problem of understanding how electricity could cross a vacuum.

- 1767: Priestley becomes Minister at Mill Hill Chapel, Leeds.
- Priestley elected a Fellow of the Royal Society as a result of HPSE
- 1769: Priestley studying powerful 'explosions' (*i.e.*, large sparks), found that small objects in powerful sparks heated and exploded; the residue was charge-free, implying the oscillatory character of the discharge.
- At Leeds, Priestley starts selling electrostatic machines for medicine and experimenters.
- Priestley shows increasing interest in gases. He discovers carbon dioxide (called by him 'fixed air') in experiments at a Leeds brewery (later states that he has found it to be a perfect conductor of electricity – an error). Names his theories 'The Doctrine of Air'.
- 1769: Watt's steam engine patent.
- 1772: Priestley publishes *The History and Present State of Discoveries Relating to Vision, Light and Colours,* a follow-up to HPSE. Some say it was 'less successful' than HPSE, others that it helped decide the wave/particle controversy.
- 1773: Priestley becomes Librarian/Amanuensis to Lord Shelbourne

Warrington Academy as extended in 1762

(Secretary of State under Prime Minister William Pitt the Elder) at Calne, Wiltshire.

- 1774: Priestley discovers 'dephlogisticated air' (*i.e.*, oxygen); Scheele also does the same in Sweden at about the same time.
- Priestley visits Paris and the Low Countries (now Holland and Belgium) with Shelburne.
- External events accelerate; Franklin returns to America 1775; Franklin's mission to France 1776–85.
- 1780: Priestley becomes the minister at New Meeting Chapel, Birmingham.
- Priestley joins the 'Lunar Society' (with James Watt, Matthew Boulton, Josiah Wedgwood, Erasmus Darwin and others) – a discussion group for leading thinkers in science.

In 1775, Volta invented the Electrophorous, a device for charging a metal plate (with an insulated handle) by induction, providing yet a new phenomenon to explain. In 1782, he proposed his famous law for capacitors, $Q = C \times T$.

Priestley pushed for bigger electrostatic generators to enable better experiments. There is some resonance here with particle accelerators in modern times. Globe designs gave way to cylinders and then to discs, which ultimately gave the best performance. Measuring instruments were developed later in the century, *e.g.*, the projection electroscope – needed for quantification. Henry Cavendish (1731–1810), a contemporary of Priestley, advanced some ideas of circuits (*e.g.*, parallel). Priestley investigated the resistance of conductors by melting wires driven by his battery of Leyden jars.

By the late 18th Century, a degree of consensus was beginning to arise, with cranks marginalised on the outside (including J. P. Marat of French Revolution assassination fame!). Electrostatics became relatively moribund from about 1790, although there were some attempts to advance theory through mathematical work done at the École Polytechnique. As in so many fields, the basics had been done and the rest was thought to be too difficult.

- 1791: Priestley and friends organise a dinner to celebrate the fall of the Bastille. An anti-radical/anti-republican mob sacks his Birmingham house, laboratory and chapel; all his laboratory equipment and many books and papers are destroyed

Medical applications and electrochemistry then became rising themes. Galvani in 1791 observed that electric shocks caused movement of frogs' legs (even when removed from the frogs) and thought he was observing a different 'animal' electricity, but others had shown that electric fish produce 'normal' electricity. The work was taken further by Volta, who found that bi-metallic joints could also drive frog legs. The voltaic pile, invented in 1800, was a non-obvious step in the development of electricity which opened up huge new field like electrolysis, steady currents, links to magnetism and electromagnetism.

Priestley's oxygen apparatus (Smithsonian Institute Washington) – including electric discharge tube

- 1791: Priestley flees to London (Hackney College and Chapel).
- 1794: Priestley emigrates to Northumberland, Pennsylvania (his sons had gone earlier) and establishes farm, substantial house and laboratory.
- 1801: *Jefferson becomes president of the USA.*
- 1804: Priestley dies and is buried at Northumberland.

By the end of 18th Century the search for explanations had declined and were replaced by mathematical descriptions. According to J. L. Heilbron (1979, p. 83), "with the large installations of the end of the 18th century, one could electrocute small animals, melt several metres of wire, electrolyse water, magnetise needles and study the chemical effects of electricity in motion".

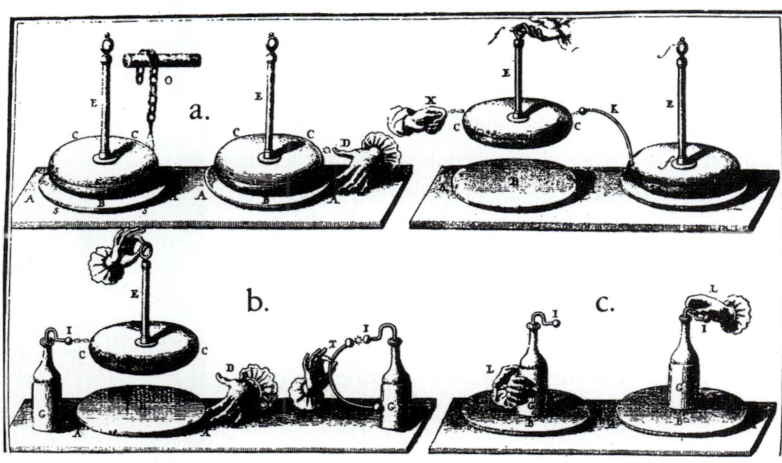

Volta's electrophorous: (a) charging by 'oscillation of the electricities', (b) charging a bottle by an electrophorous, (c) charging an electrophore by a bottle

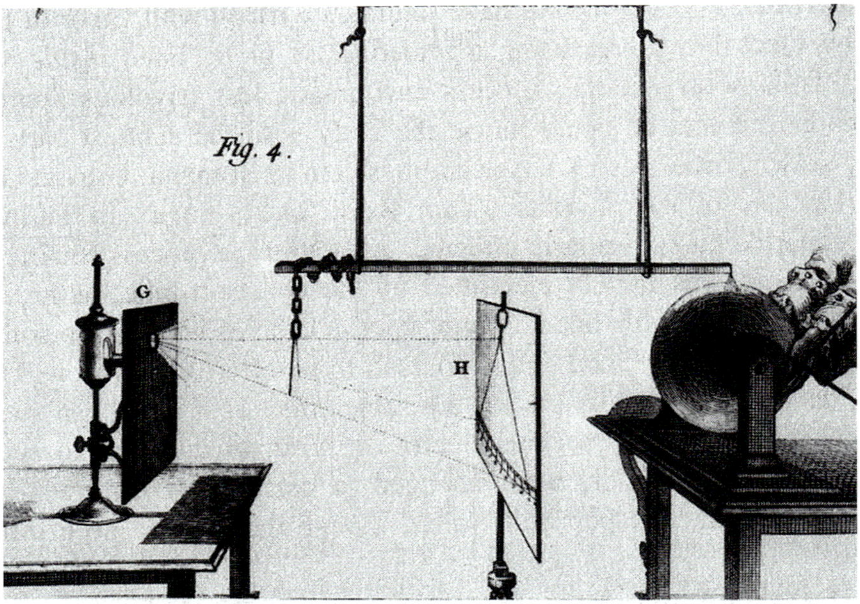

Nollet's electroscope: the lamp at G images the threads from the prime conductor on the screen H

19th Century

The concept of electric potential was established in 1811 by Siméon-Denis Poisson (1781–1840). There was still no connection made between electricity and magnetism until 1820, when Hans Christian Oersted (1777–1851) proposed that the two phenomena were linked; Faraday's Law of electromagnetism did not emerge until 1831. 'Galvanism' was still regarded as a separate phenomenon in a 1823 textbook.

The existence of coulombic forces in matter was not considered until much later; this required a linkage of the concept of vacuum spaces in atomistic matter to that of a macroscopic vacuum. There were some hints of thoughts along these lines in eighteenth century but no more than that.

According to Heilbron (1979), "The chief moral ... may be that, when confronted with a choice between a qualitative model, deemed intelligible, and an exact description lacking clear physical foundations, the leading physicists of the Enlightenment preferred exactness". One might compare this with Paul Dirac's famous statement in his *Principles of Quantum Mechanics* of 1930, "The only object of theoretical physics is to calculate results that can be compared with experiment, and it is quite unnecessary that any satisfactory description of the whole course of the phenomena should be given".

The Priestley monument erected by the American Chemical Society

References and Sources

P. S. Excell, 'Sir Edward Appleton and Joseph Priestley: Two Giants of Electrical Science', J. Atmos. Terr. Phys., 1994, 56, 693–704.

F. W. Gibbs, *Joseph Priestley*, Nelson, London, 1965.

J. L. Heilbron, *Electricity in the 17th and 18th Centuries*, Univ. California Press, Berkeley, 1979.

T. S. Kuhn, *The Structure of Scientific Revolutions*, Univ. Chicago Press, Chicago, 1970.

J. Priestley, *The History and Present State of Electricity, with Original Experiments*, Dodsley, Johnson, Davenport and Cadell, London, 1766.

J. Priestley, *The Memoirs of Dr Joseph Priestley, to the Year 1795*, J. Johnson, London 1806.

R. E. Schofield (ed.), *A Scientific Autobiography of Joseph Priestley (1733–1804)*, M. I. T. Press, Cambridge, USA, 1966.

NOTES

1. Significant events in Joseph Priestley's life are placed in context with the historical development of electricity by insertions of this kind
2. Several pages from HPSE and pictures of the equipment used by Priestley are given in the Supplement to this paper.

SUPPLEMENT – PAGES FROM THE HISTORY AND PRESENT STATE OF ELECTRICITY *(HPSE)* *AND PICTURES OF PRIESTLEY STYLE ELECTRICAL APPARATUS*

THE

HISTORY

AND

PRESENT STATE

OF

ELECTRICITY,

WITH

ORIGINAL EXPERIMENTS,

By JOSEPH PRIESTLEY, LL.D. F.R.S.

THE THIRD EDITION,

CORRECTED AND ENLARGED.

Causa latet, vis est notissima. OVID.

VOL. I.

LONDON,

Printed for C. BATHURST, and T. LOWNDES, in Fleet-Street; J. RIVINGTON, and J. JOHNSON, in St. Paul's Church-Yard; S. CROWDER, G. ROBINSON, and R. BALDWIN, in Pater-noster Row; T. BECKET, and T. CADELL, in the Strand.

MDCCLXXV.

THE

CONTENTS.

PART I.

THE HISTORY OF ELECTRICITY.

PERIOD I.

PERIOD II.

PERIOD III.

PERIOD IV.

PERIOD V.

PERIOD VI.

PERIOD VII.

VOL. I. c PE-

PERIOD VIII.

PERIOD IX.

PERIOD X.

Sec.

SEC.

THE

CONTENTS.

PART II.

PART III.

PART IV.

A 2 III.

PART V.

PART

PART VI.

PART VII.

PART VIII.

SEC.

SEC.

XII. Ob-

THE

HISTORY

AND PRESENT STATE OF

ELECTRICITY.

PART II.

A

SERIES OF PROPOSITIONS,

COMPRISING ALL THE

GENERAL PROPERTIES OF ELECTRICIY,

AFTER tracing, at large, the progreſs of all the diſcoveries relating to electricity, and giving an hiſtorical account of them, in the order in which they were made; it will, probably, be no diſagreeable repetition, if I give, at the cloſe of it, a SERIES OF PROPOSITIONS, compriſing all the general properties of electricity, drawn up in as ſuccinct a manner as poſſible. And, notwithſtanding the large detail which has been made, it will be found, that a few propoſitions are ſufficient to compriſe almoſt all that we know of the ſubject.

THIS circumſtance may be regarded as a demonſtration of the real progreſs that has been

VOL. II. B made

The Science Museum (London) Generator

HPSE: Globe-type electrostatic generator – as sold by Priestley for medical use

Priestley globe-type electrostatic generators – Smithsonian Institute, Washington (*top left*) and Science Museum, London (*top right*)

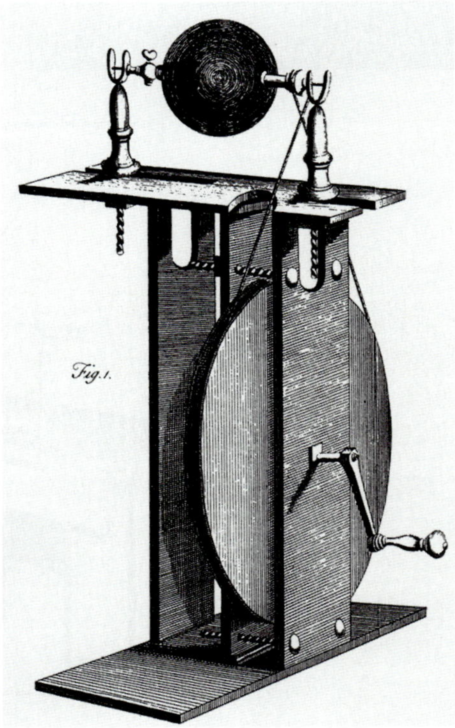

HPSE: Globe-type generator with spark discharge experimental equipment (*left*). Note use of wire.

HPSE: Variant of globe-type (*right*)

HPSE: Paralleled electrostatic generators

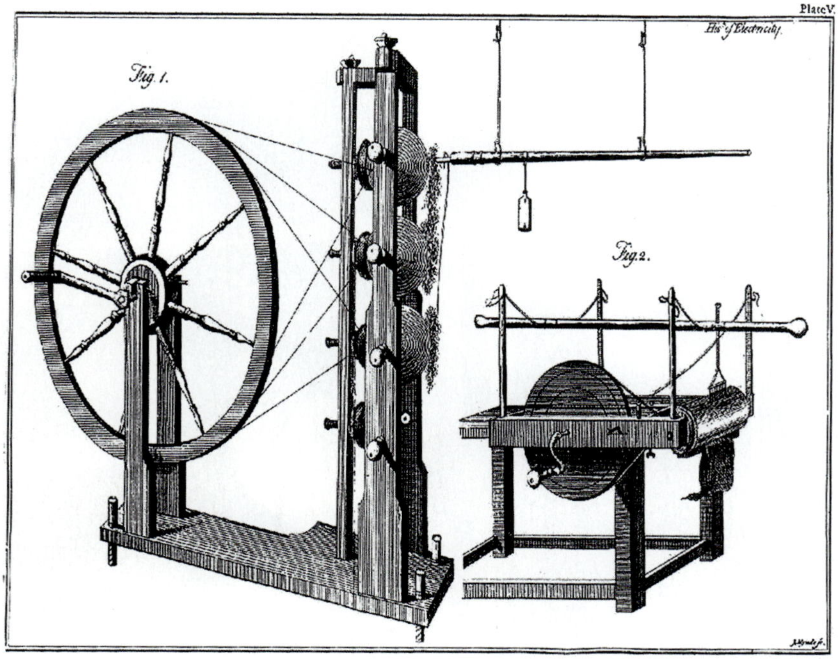

HPSE: Vertical-axis cylinder-type electrostatic generator

Cylinder-type electrostatic generator (Science Museum)

(top left) Cylinder-type electrostatic generator

(top right) Science Museum: Disc-type electrostatic generator

HPSE: Arrangement for transmitting electricity

HPSE: Another arrangement for transmitting electricity

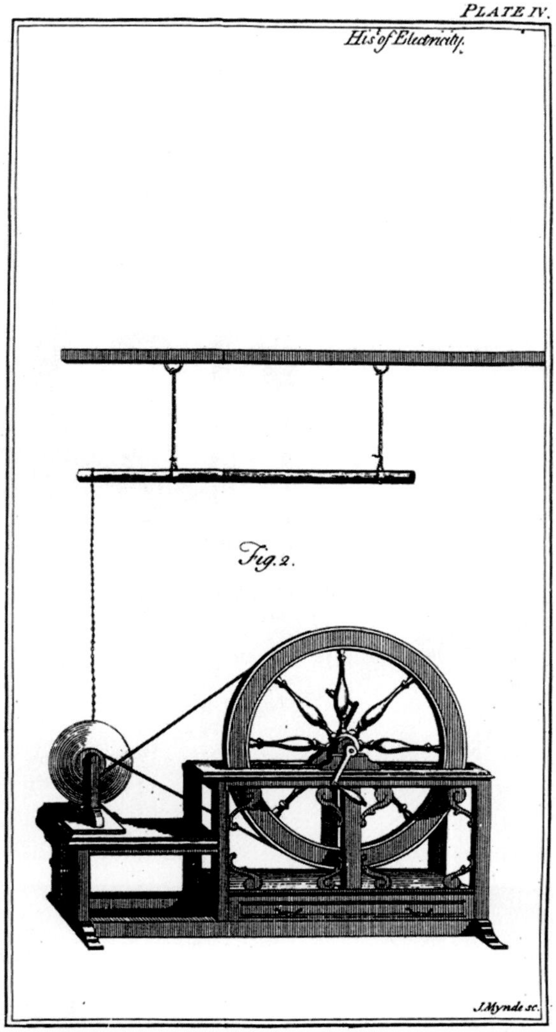

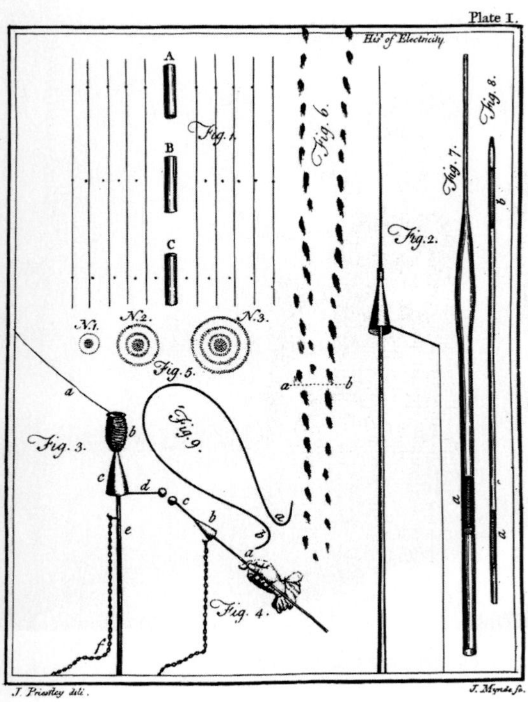

HPSE: Theoretical ideas and spark discharge equipment

HPSE: Electrostatic experiments; Rear: Leyden jar capacitor; Left: Automatic bells operated by electrostatic force and discharge

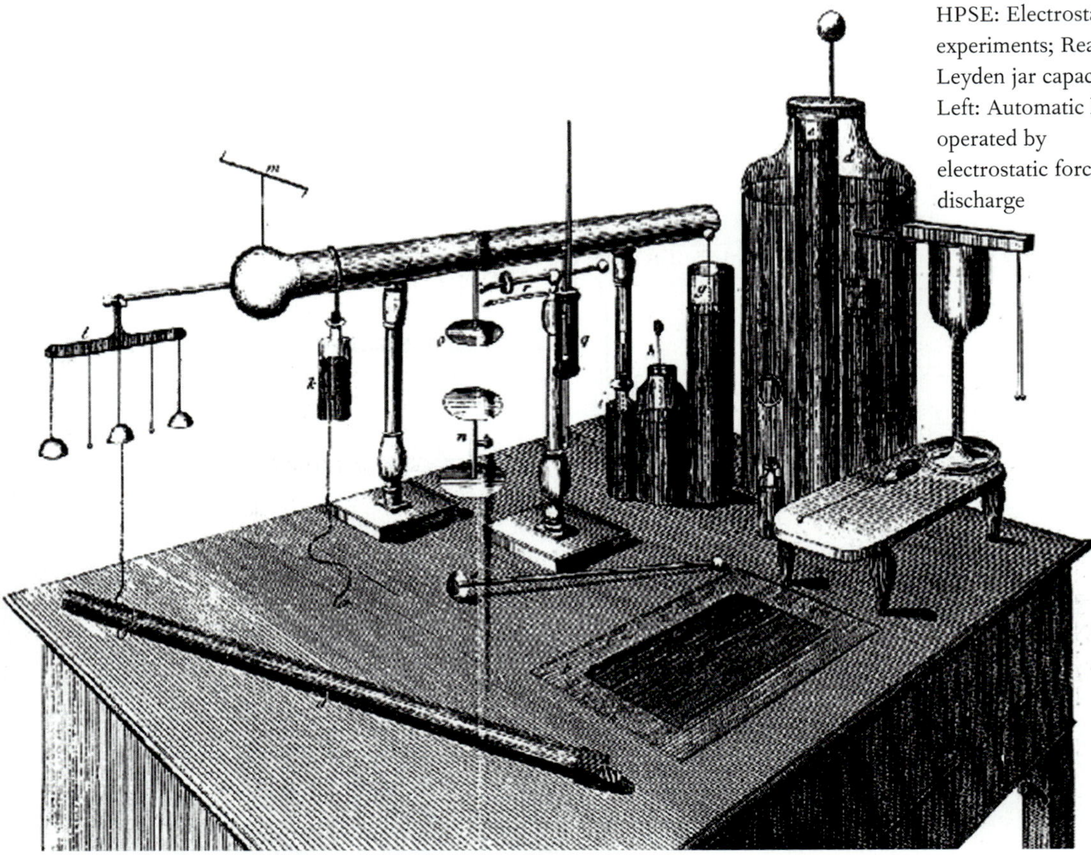

Science Museum: Cylinder generator; Leyden jar; automatic bells; etc.

HPSE: Large bank of Leyden jars

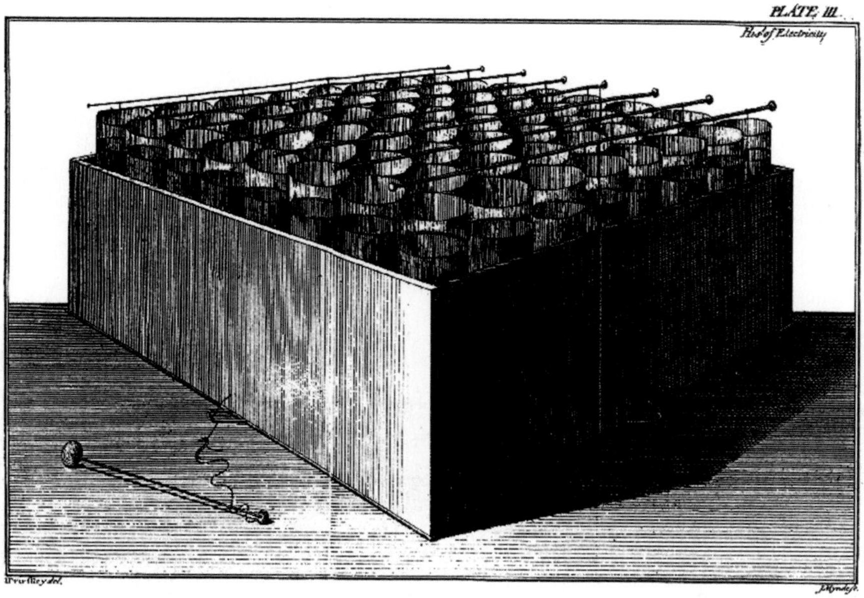

held, the higher charge it received; the wire of the phial, which communicated with the inſide coating, being farther removed from the influence of the electricity of the cup.

May we not infer from this experiment, that the attraction of electricity is ſubject to the ſame laws with that of gravitation, and is therefore according to the ſquares of the diſtances; ſince it is eaſily demonſtrated, that were the earth in the form of a ſhell, a body in the inſide of it would not be attracted to one ſide more than another.

Doth it not follow from the experiments of the balls, compared with thoſe with the phial, that no body can receive electricity in one place, unleſs an opportunity be given for its parting with it in another; at leaſt, that a quantity muſt be repelled from any particular part before any more can enter, ſince a ſmall body can no more receive electricity when all its ſides are equally expoſed to the action of an electrified body, than a phial can be charged when both its coatings are equally expoſed to the ſame electricity?

Do not theſe experiments, likewiſe, favour the hypotheſis of S. Beccaria, that there is no electrical attraction without a communication of electricity?

Mr. Lullin alſo made ſeveral of theſe experiments with an electrified hollow veſſel.

Almost the last page, and Priestley spots what is now known as Coulomb's Law of electrostatic force fields; beginning of first full paragraph

XV. Experiments with an ELECTRIFIED CUP.

I shall cloſe the account of my experiments with a ſmall ſet, in which, as well as in the laſt, I have little to boaſt beſides the honour of following the inſtructions of Dr. Franklin. He informed me, that he had found cork balls to be wholly unaffected by the electricity of a metal cup, within which they were held; and he deſired me to repeat and aſcertain the fact, giving me leave to make it public.

Accordingly, December the 21ſt. I electrified a tin quart veſſel, ſtanding upon a ſtool of baked wood; and obſerved, that a pair of pith balls, inſulated by being faſtened to the end of a ſtick of glaſs, and hanging intirely within the cup, ſo that no part of the threads were above the mouth of it, remained juſt where they were placed, without being in the leaſt affected by the electricity; but that, if a finger, or any conducting ſubſtance communicating with the earth, touched them, or was even preſented towards them, near the mouth of the cup, they immediately ſeparated, being attracted to the ſides; as they alſo were in raiſing them up, the moment that the threads appeared above the mouth of the cup.

If the balls had hung in the cup a conſiderable time without touching it, and they were

were taken out immediately after the electricity of the cup was diſcharged, they were found to have acquired no degree of electricity.

If they had touched any part of the cup, though they ſhowed no electricity while they were within it; yet, upon being taken out, they appeared to have acquired ſome; which was more if they had touched a part near the edge of the cup, leſs if they had touched any part more remote from the edge, and leaſt of all if they had touched the bottom only. If they had firſt touched the ſide near the top, and then the bottom, they came out with that ſmall degree of electricity which they would have acquired, if they had touched the bottom only.

In any caſe, if the balls were taken out while the cup remained electrified, they neceſſarily acquired ſome degree of electricity, in paſſing the mouth of the cup.

To purſue this experiment a little farther, I took a ſmall coated phial, ſuch as is repreſented upon the ſtool [*c* Pl. II.] and obſerved, that when I held it by the wire, within the electrified cup, it acquired no charge, the electricity of the cup affecting both the inſide and outſide coating alike. If the external coating touched the bottom of the cup, the phial received a very ſmall charge. If it was made to touch the ſide, it acquired a greater charge; and the nearer to the top it was held,

held, the higher charge it received; the wire of the phial, which communicated with the inſide coating, being farther removed from the influence of the electricity of the cup.

MAY we not infer from this experiment, that the attraction of electricity is ſubject to the ſame laws with that of gravitation, and is therefore according to the ſquares of the diſtances; ſince it is eaſily demonſtrated, that were the earth in the form of a ſhell, a body in the inſide of it would not be attracted to one ſide more than another.

DOTH it not follow from the experiments of the balls, compared with thoſe with the phial, that no body can receive electricity in one place, unleſs an opportunity be given for its parting with it in another; at leaſt, that a quantity muſt be repelled from any particular part before any more can enter, ſince a ſmall body can no more receive electricity when all its ſides are equally expoſed to the action of an electrified body, than a phial can be charged when both its coatings are equally expoſed to the ſame electricity?

Do not theſe experiments, likewiſe, favour the hypotheſis of S. Beccaria, that there is no electrical attraction without a communication of electricity?

MR. LULLIN alſo made ſeveral of theſe experiments with an electrified hollow veſſel.

ſel. He obſerved, that if an electrified cork ball, ſuſpended by a ſilken ſtring, was let down into the veſſel, and touched the bottom, it left all its electricity behind it. He alſo made the experiment with a glaſs veſſel, coated at the bottom and charged, with the ſame event. It alſo made no difference whatever was the form, or ſize of the body let down into it, provided it was one third leſs than the depth of the veſſel; but if, when it touched the bottom, it, at the ſame time, reached the top, or came near the top of the veſſel, it acquired electricity; and a conſiderable degree, if it exceeded the top. The form of the veſſel made no difference in his experiments, nor did it make any whether the veſſel was intire or perforated. A wire net anſwered perfectly well. Theſe experiments, Mr. Lullin imagines, clearly prove Nollet's doctrine of the conſtant motion of electric atmoſpheres: for he thinks, that this free motion on which electrification depends, is prevented from taking place within the veſſel, by the contrary tendency of the oppoſite ſides *.

DR. FRANKLIN, in the laſt edition of his Letters, p. 326, ſays, that poſſibly the mutual repulſion of the oppoſite ſides of the

* Diſſertatio phyſica, p. 38.

electrified

376 MISCELLANEOUS, &c,

electrified cup, may prevent the accumulation of the electric atmosphere upon them, and occasion it to stand chiefly on the outside. But he recommends it to the farther examination of the curious.

A CATA-

A CATALOGUE OF BOOKS WRITTEN ON THE SUBJECT OF ELECTRICITY, EXCLUSIVE OF PAPERS IN BOOKS OF PHILOSOPHICAL TRANSACTIONS, AND OTHER MISCELLANEOUS WORKS; DISTINGUISHING [BY ASTERISMS] THOSE WHICH THE AUTHOR HAD SEEN, AND MADE USE OF IN COMPILING THIS WORK.

* GILBERT de Magnete Magneticisque corporibus, 1600, London, folio.

* *Otto de Guericke*'s experimenta nova Magdeburgica, 1672, Amsterdam, folio.

* *Hauksbee*'s physico mechanical experiments, 1709, 1719, London, octavo.

G. M. Bose's oratio inauguralis de electricitate, 2 parts, 1738, Wittemburg. Gralath's bibliothek.

J. Martenson's dissertatio de electricitate, 1740, 1742, Upsal, quarto, Gralath's electritche bibliothek.

* *Desaguliers*' dissertation concerning electricity, 1742, London.

C. A. Hausen novi profectus in historia electricitatis, Leipsic, 1743.

C. G. Kratzenstein's abhandlung von der nutzen der electricitat in der artzney wissenschafft, 2d. edit. 1745, Halle. Gralath's bibliothek.

A. G. Rosenberg's versuch einer erklarung von den ursachen der electricitat, 1745. Breslau. Gralath's bibliothek.

Nollet's conjettures sur les causes de l' elettricité des corps, Paris, 1745. Wilson, p. 12.

——— essai sur l' electricité des corps, 1746, 1754. Paris.

* *Waitz*'s abhandlung von der electricitat und deren ursachen, together with two other essays in High Dutch, and one in French, all written for the prize proposed by the academy at Berlin, 1745, Berlin, quarto.

* *Wilson*'s essay towards an explication of the phenomena of electricity deduced from the ether of *Sir Isaac* Newton, 1746, London, octavo.

* *Watson*'s experiments and observations, tending to illustrate the nature and properties of electricity, 3d edit. 1746, London, octavo.

* ——— sequel to the experiments and observations, tending to illustrate the nature and properties of electricity, 1746, 1747. London, octavo.

* *Freke*'s essay to show the cause of electricity, 1746, London, octavo.

Kratzenstein's theoria electricitatis, more geometrico explicata, Hall, 1746.

J. Pideris's dissertatio inauguralis de electricitate, 1745, Marburgh. Gralath's bibliothek.

* *Martin*'s essay on electricity, 1748, Bath, octavo.

Bose's recherches sur la cause, et sur la veritable theorie de l' electricité. 146

VOL. II. Cc Con-

FROM BOYLE TO PRIESTLEY AND THE BIRTH OF CHEMISTRY

John Maguire

John Maguire
Air Force Research Laboratory
Dayton, Ohio

In the foyer of the Chemical Heritage Foundation building in Philadelphia hang the portraits of Robert Boyle (by Johann Kerseboom, 1689) and Joseph Priestley (by Ozias Humphrey 1810).

[Visitors to the Chemical Heritage Foundation will note that the Kerseboom portrait was purchased from the Estate of Viscount Shannon.]

The symbolism could hardly resonate more strongly in that the period from the birth of Robert Boyle (1627) to the death of Joseph Priestley (1804) marked the birth of modern chemistry. This was the time when Aristotelian philosophy of the ancients was gradually being displaced by Baconian observation and empiricism and when the foundation of the modern scientific method, including the interplay between theory and experiment, was becoming established on firm foundations. At the start of the seventeenth century chemical science was significantly influenced by the alchemical notions of the adepts, by mysticism and by the occult, but by the end of the eighteenth century Priestley had discovered oxygen, the notion of chemical reaction and oxidation were well known and, most importantly, the first triumph of physical chemistry namely the Law of Conservation of Matter had been demonstrated unambiguously.

In this short presentation it would be foolhardy to attempt to cover over two hundred years of chemical history. Our purpose today, at this bicentennial

Robert Boyle (*left*) by Johan Kerseboom (1689) Joseph Priestley (*right*) by Ozias Humphrey (1810)

Lady Catherine Boyle (*left*)

Boyle Memorial, 1632 – St. Patrick's Cathedral, Dublin (*right*)

celebration of Joseph Priestley, will be far more limited. I shall attempt to tell a much simpler story and sketch how the work of Robert Boyle established a methodology, a direction, and a scientific culture that was further developed by talented investigators who, in turn, set the stage for the discoveries of Joseph Priestley and the establishment of the law of conservation of matter.

Our story starts on January 25th, 1627 when Lady Catherine Boyle gave birth to a son Robert known then as Robin at his home in Lismore Castle, near Waterford, Ireland. Robert was the fourteenth child and seventh son of the First Earl of Cork, Richard Boyle. His mother would die shortly after the birth of her next child, a daughter Margaret who did not survive to adulthood. The Boyle Memorial in St. Patrick's cathedral in Dublin was constructed in honour of Lady Catherine with baby Robin depicted on the pedestal in the middle of the bottom tier.

Lismore Castle

Earl of Cork as an Adventurer

Richard Boyle, the first Earl of Cork had been educated at Cambridge and came to Ireland in 1588 at age 22 with a small amount of money, some fine clothes and a diamond ring; all of which he appears to have used to significant advantage. He initially married (1595) Joan Apsley, the daughter of a wealthy Limerick merchant. She died in childbirth in 1599 and although the Earl did not erect a monument to her memory he did continue to draw on her dowry (five hundred pounds per year) until at least 1632. The Earl seems to have taken the Boyle family motto "God's providence is my inheritance" in an almost literal sense. He married Catherine Fenton, the daughter of an influential Dublin family, and again the heiress to substantial fortune, on the 25th July 1603, the same day he was knighted in Dublin castle.

The Earl was quite an operator. The kindest that might be said of his business dealing is that he 'leveraged' a series of land deals that amounted to confiscations. He was imprisoned for the forgery of titles but managed to secure a Royal pardon by means that are not entirely clear but we may assume that he exercised what in America is now termed a substantial 'lobby'. The Earl was clearly not a man to be trifled with. He was, of course, unpopular among the native Irish. Oliver Cromwell remarked, some years after the Earl's death, that if there had been an Earl of Cork in every Province, Ireland could never have risen in rebellion. Even among his peers the Earl was not universally appreciated. For example, Archbishop William Laud wrote to Richard Boyle's prosecutor, Thomas Wentworth,

> no physic better than a vomit if it be given in time, and therefore you have taken a very judicious course to administer one so early to my Lord of Cork. I hope it will do him good

He was the ultimate political survivor and even beat a charge of treason (a truly life-threatening event) for holding discussions with the Spanish at the time of the Armada. However, giving evidence at his trial in person, Queen Elizabeth I declared famously,

> By God's death, these are but inventions against the young man … a man fit to be employed by ourselves

The Earl beat the wrap and was indeed promptly employed by the Queen, being appointed first Clerk of Munster in 1600 and by 1640 he was Lord Chief Sheriff of Ireland and one of the richest men, if not *the* richest man, in the United Kingdom. He also lived, albeit in exceedingly perilous times, to see both Laud and Wentworth hanged after he gave evidence against them. Clearly, great caution was the order of the day if you chose to cross the Earl of Cork!

Robert Boyle was born into this family of great wealth, power, prestige and privilege on 25th January 1627. He was educated at home in Lismore Castle and at the age of eight was sent with his brother Francis to school in Eton. He had a good knowledge of both Latin and Greek before going to Eton but was bullied and beaten so badly that he forgot his languages and had to relearn them later. He also developed a stutter and was taken out of Eton in 1638. In his autobiography he notes that he was his father's favourite and preferred academic pursuits to playing games.

In 1639 the boys left England with their tutor, a Mr. Marcombes, for the grand tour of Europe and spent extended time in Italy, France and Switzerland. He met many of the leading scientists and mathematicians in Europe and was in Florence at the start of 1642 when the great old man, Galileo, died. He was well aware of the work of Galileo and the manometers of Torricelli, Galileo's secretary at the time.

It was also in Italy that Boyle reports in his autobiography one of his very few sexual encounters. He visited an Italian brothel with Marcombes and his brother Francis but insists that he went merely out of youthful curiosity and did not engage in the business of the establishment.

> Nor did he sometimes scruple, in his Governor's company, to visit the famousest Bordellos; whither resorting out of bare curiosity, he retained there an unblemish't Chastity, and still return'd thence as honest as he went thither
>
> *Boyle, Winter 1641–42*

Boyle, however, seems curiously oblivious to the fact that in reporting this matter he might in any way implicate both his teacher and brother. One might reasonably enquire as to what they were doing while young Robin was merely satisfying his boyish curiosity! Later in the trip he reports being accosted by two priests but again made a lucky escape,

> *But though P. were no Fewell for forbidden Flames, he prov'd the Object of unnaturall ones. For being at that Time not above 15, & the Cares of the World having not yet faded a Complexion naturally fresh enuf; he was somewhat rudely presst by the Preposterous Courtship of 2 Fryers, whose Lust makes no distinction of the Sexes ... forc't a scape from these gown'd Sodomites.*

A portrait of Boyle at around then age fifteen shows that he was indeed a young man of exceedingly fair complexion and you can see from a much later portrait that the worries of the world had not dulled his fair complexion even well into middle age.

The return to England was delayed for two years when the party became stranded in France. Two things were going on. First, the Irish had risen (yet again) in rebellion and second, civil war had broken out in England. The boys ran out of money and since they did not have the convenience of a modern banking system were obliged to live on the earning of their tutor. The Earl managed to send them some jewellery to sell but this was a risky and

Robert Boyle at about 15

Boyle in middle age. Wigs were in high fashion

unreliable activity, oftentimes presenting more temptation than the messenger could bear. The family fortune was in dire jeopardy when the Earl wrote,

> out of some meet port in France to land either in Dublin, Corke, or Youghall, (for all the other cities and sea townes are possessed by the enemy), or else my two sonnes travaile into Holland, and putt themselves into entertaynement under the service and conduct of the Prince of Orange; for they must henceforward maintayne themselves by such entertaynements as they gett in warres
>
> Earl of Cork March 1642

In the event Francis returned to Ireland and fought at the battle of Liscarroll on the 20th August 1642. In that battle their brother Lewis Boyle (Lord Kynalmeaky) was killed by a shot from the Irish musketeers "from behind bushes and ditches but his horse was brought off by his brother Francis Boyle (later Viscount Shannon)" (*opposite*) Title page of the Second Edition of "New Experiments Physico-Mechanical Touching the Spring of the Air" (Copy presented by Boyle to John Evelyn)

During his time on the continent of Europe, Robert Boyle learned mathematics and the new science that had been developed largely in the Jesuit schools. He also, of course, came to appreciate the power of observational science and came to reject absolutely the doctrine of informed ignorance of which Galileo had fallen foul but has since been pardoned on October 14th, 2001.

Boyle eventually returned to England in mid-1644 and set up a laboratory at Stalbridge manor, a large country house in Dorset that he had inherited from his father. Boyle set about some initial experiments but was mostly concerned with religious philosophy. There was certainly no hint at this time of any great scientific promise. He returned to Ireland briefly (one year from June 1652 and for eight months in October 1653) after Cromwell had very efficiently subdued the rebellion, notably at the city of Drogheda, and secured his fathers estates with a generous settlement. That things had gone well for the English was obviously of immense significance to both the security and prosperity of England. But it might also be mentioned that the event also had an impact on chemistry. Robert Boyle would not have to go into business in Holland but could provide secure long-term funding for curiosity-driven investigations in what was the first modern 'research group'.

Boyle moved to Oxford in 1654 and began by far the most productive part of his career. It was during this period that he carried out the celebrated experiments on gases. The design of the apparatus, with all manner of joints, valves, and fixtures for the introduction and retrieval of samples was largely the work of Robert Hooke. These were first published in 1662 in "A Defence of the

(*left*) One of the many likenesses of Robert Hooke

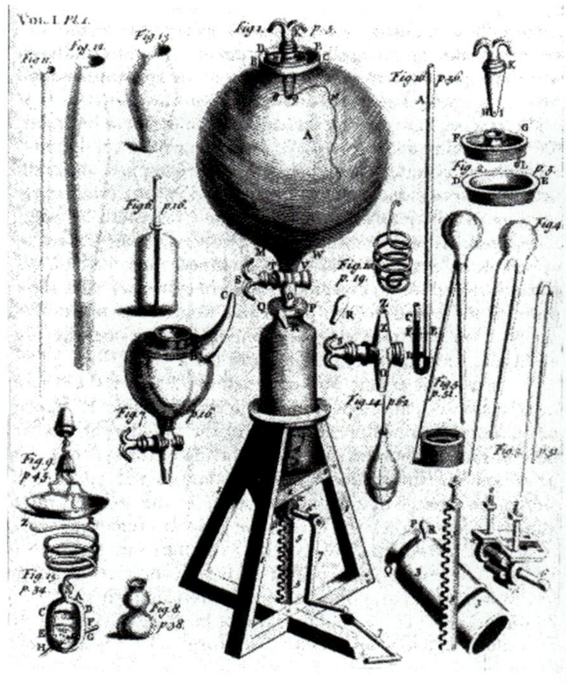

(*right*)The Boyle-Hooke vacuum pump with accessories

NEW
EXPERIMENTS
PHYSICO-MECHANICAL,

Touching

The SPRING of the AIR, and its EFFECTS,

(Made, for the most part, in a New

Pneumatical Engine)

Written by way of LETTER

To the Right Honorable *CHARLES* Lord

Vicount of *DUNGARVAN,*

Eldest Son to the EARL of *CORKE.*

By the Honorable *Robert Boyle* Esq;

OXFORD: Printed by *H. Hall*, Printer to the University,

for *Tho: Robinson*. 1662.

Doctrine Touching the Spring and Weight of the Air" appended to the Second Edition of the *New Experiments Physico-Mechanical, Touching The Spring of the Air, and its Effects (made for the most part, in a New Pneumatical Engine)*. The book, with its characteristically prolix title, was dedicated to his brother Lord Dungarvan. In this work, Boyle addresses the effects of temperature on gases and describes what would become known as Boyle's law although Boyle never used the term. Since the thermometer had not been invented, he could not investigate in any quantitative fashion what would later be known as Charles's Law.

However, the effects of pressure at constant temperature were certainly captured fairly accurately but were never stated as a generalization, or law, for Boyle would have demurred on that point. He was much too cautious to assume more general applicability than the measurements dictated (see Table below);

> all that I shall now urge being, that however the trial already made sufficiently proves the main thing, for which I here allege it; since by it, it is evident, that as common air, when reduced to half its wonted extent, obtained near about twice as forcible a spring as it had before; so this thus comprest air being further thrust into half this narrow room, obtained thereby a spring about as strong again as it last had, and consequently four times as strong as that of the common air.

This is the first statement of what is now called Boyle's Law in England and in the USA. Note from the table that the data are in the form of fractions and that there is, of course, no straight line plot in the original publication.

A Table of the Condensation of the Air

A	A	B	C	D	E
48	12	00	Added to 29⅛ makes	29 2/16	29 2/16
46	11½	01 7/16		30 9/16	30 6/16
44	11	02 13/16		31 15/16	31 12/16
42	10½	04 6/16		33 8/16	33 1/7
40	10	06 3/16		35 5/16	35 --
38	9½	07 14/16		37 --	36 15/19
36	9	10 2/16		39 5/16	38 7/8
34	8½	12 8/16		41 10/16	41 2/17
32	8	15 1/16		44 3/16	43 11/16
30	7½	17 15/16		47 1/16	46 3/5
28	7	21 3/16		50 5/16	50 --
26	6½	25 3/16		54 5/16	53 10/13
24	6	29 11/16		58 13/16	58 2/8
23	5¾	32 3/16		61 5/16	60 18/23
22	5½	34 15/16		64 1/16	63 6/11
21	5¼	37 15/16		67 1/16	66 4/7
20	5	41 9/16		70 11/16	70 --
19	4¾	45 --		74 2/16	73 11/19
18	4½	48 12/16		77 14/16	77 2/3
17	4¼	53 11/16		82 12/16	82 4/17
16	4	58 2/16		87 14/16	87 3/8
15	3¾	63 15/16		93 1/16	93 1/5
14	3½	71 5/16		100 7/16	99 6/7
13	3¼	78 11/16		107 13/16	107 7/13
12	3	88 7/16		117 9/16	116 4/8

AA. The number of equal spaces in the shorter leg, that contained the same parcel of Air diversly extended.
B. The height of the Mercurial Cylinder in the longer leg, that compress'd the Air into those dimensions.
C. The height of a Mercurial Cylinder that counterbalanc'd the pressure of the Atmosphere.
D. The Aggregate of the two last Columns *B* and *C*, exhibiting the pressure sustained by the included Air.
E. What that pressure should be according to the *Hypothesis*, that supposes the pressures and expansions to be in reciprocal proportion.

Boyle's original data (for relative heights of mercury in the J-tube pictured overleaf). Note the use of complicated fractions

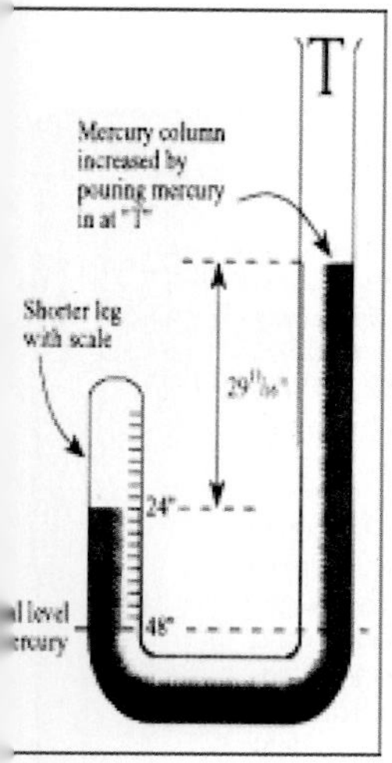

A schematic of the J-shaped tube used in the experiments on the compression of gases

In France, Boyle's law is known as Mariott's law. Edme Marriott was working on gases at about the same time as Boyle but published later. However, Mariott did call the inverse relationship a "law" which Boyle never did. Similarly, Richard Townley, who worked with Boyle and came from a prominent Catholic family in the North of England had noted the variations in manometer height with altitude and extrapolated the results to pressures lower than one atmosphere. Ralf Gratorix certainly helped build the apparatus, which was designed largely by Hooke who almost certainly was responsible for the analysis and preparation of the table of results.

It was really during the time in Oxford that the work of the Boyle group really flourished, supported by an able group of investigators who were, no doubt, happy to enjoy his patronage and proximity to the rich and powerful. On this Fourth of July celebration of Joseph Priestley, it is appropriate to mention the American member of the Boyle group, Robert Starkey. Starkey had graduated from Harvard in 1643–6 and emigrated to England with his wife. He was heavily involved with Boyle and is reputed to have taught Boyle the essentials of quantitative chemistry, which Boyle embraced wholeheartedly. As has been pointed out recently, (*Alchemy Tried in the Fire: Starkey, Boyle, and the Fate of Helmontian Chymistry*, William R. Newman and Lawrence M. Principe, University of Chicago Press, 2002), Starkey's work and notebooks reveal a fundamentally modern approach to science, with review of the literature, conjecture and hypothesis followed by experimentation and conclusions. Starkey went bankrupt and was imprisoned (apparently Boyle's patronage had limits!) for two years and while the interaction appears to have continued for some time after his release Starkey died in the plague in 1665 though the official cause of death was apparently "immoderate ingurgitation of dull flat Small beer".

Boyle's *The Sceptical Chymist* was published in 1661 and was to have great influence in the gradual rejection of alchemical principles and the establishment of chemical science. On one page Boyle anticipates, for he had no evidence, molecules, stereochemistry, and the kinetic theory of gases.

Boyle also conducted a number of experiments that would have significant bearing on the work of Joseph Priestley. For example, it had been known since the publication of Agricola's *De Rera Metallica* that 118 lb of tin could be obtained from 150 lb of ore. Boyle had read a report from the continent of Europe that claimed that when tin was heated it formed a calx (oxide) with a gain in weight. He thought that the apparent gain in weight probably came from ash from the fire finding its way into the molten tin. He repeated the experiment in a partially evacuated flask weighing the crucible before and after the heating operation. Boyle found that the tin did indeed gain weight but had no idea where the increase in weight came from. While Boyle weighed the crucible before and after heating, he did not think to weigh the sealed flask before and after heating. Had he done so he would have come tantalizingly close to Lavoisier's major discovery of more that 100 years later.

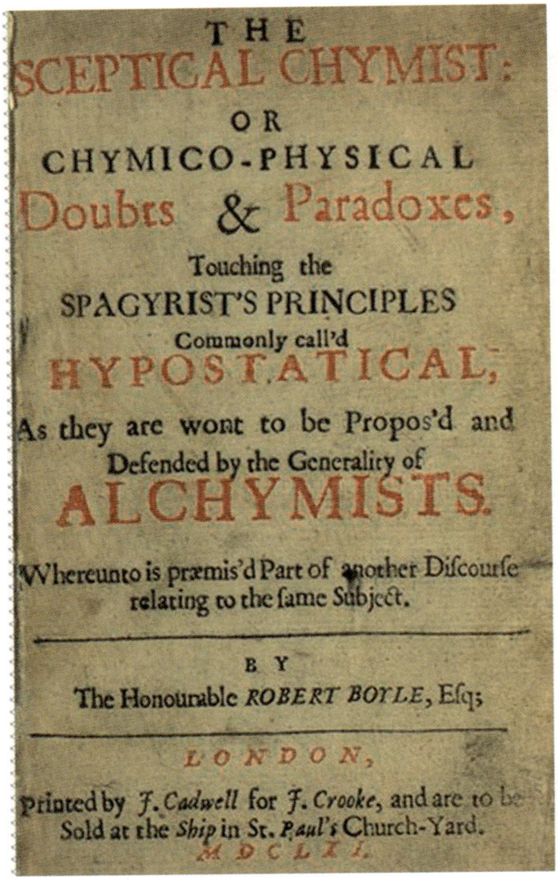

THE
SCEPTICAL CHYMIST:
OR
CHYMICO-PHYSICAL
Doubts & Paradoxes,
Touching the
SPAGYRIST'S PRINCIPLES
Commonly call'd
HYPOSTATICAL;
As they are wont to be Propos'd and
Defended by the Generality of
ALCHYMISTS.
Whereunto is præmis'd Part of another Discourse
relating to the same Subject.

BY
The Honourable *ROBERT BOYLE*, Esq;

LONDON,
Printed by *J. Cadwell* for *J. Crooke*, and are to be
Sold at the *Ship* in St. *Paul's* Church-Yard.
MDCLXI.

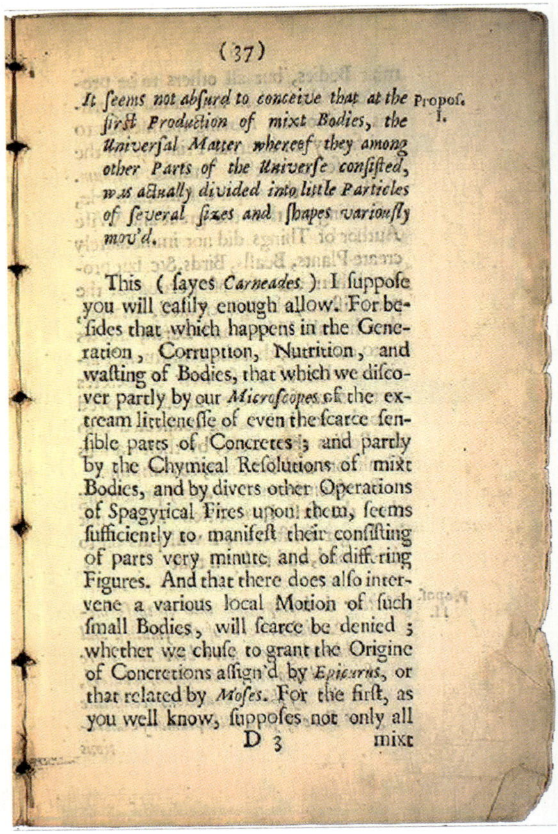

(37)

It seems not absurd to conceive that at the first Production of mixt Bodies, the Universal Matter whereof they among other Parts of the Universe consisted, was actually divided into little Particles of several sizes and shapes variously mov'd. Propos. I.

This (sayes *Carneades*) I suppose you will easily enough allow. For besides that which happens in the Generation, Corruption, Nutrition, and wasting of Bodies, that which we discover partly by our *Microscopes* of the extream littleness of even the scarce sensible parts of Concretes; and partly by the Chymical Resolutions of mixt Bodies, and by divers other Operations of Spagyrical Fires upon them, seems sufficiently to manifest their consisting of parts very minute, and of differing Figures. And that there does also intervene a various local Motion of such small Bodies, will scarce be denied; whether we chuse to grant the Origine of Concretions assign'd by *Epicurus*, or that related by *Moses*. For the first, as you well know, supposes not only all

D 3 mixt

Title page of "The Sceptical Chymist" (1661) (*left*)

A single page speculates on molecules, stereochemistry and the kinetic theory

Boyle worked on a huge number of experiments and made many speculations,

> I have often suspected that air is not, as many imagine, a simple and elementary body, but a confused aggregate of effluviums constituting by their minuteness and various motions, one great mass of fluid matter.

He also described the generation of impure hydrogen,

> A very volatile and saline piercing liquor (hydrochloric acid) being dropped upon filings of steel, the mixture grew hot. And there was emitted, out of the vial that the steel was contained in, very fetid steams. These would kindle at the flame of a candle and continue to burn a good while. A good name for this would be inflammable air.

Robert Boyle died in 1691, one week after the death of his sister, Katherine, with whom he lived in later life and who, thirteen years his senior, had played the role of a surrogate mother after the death of Lady Catherine. A devoutly religious man, he left an endowed "Boyle Lecture" for the conversion of Jews, America Indians, Irish, and Moslems.

Boyle's approach, while ultimately successful did not take hold immediately. Johann Kunckel (alias Johann von Löwernstern-Kunckel, 1630–1702), whose father was alchemist to the court of Holstein, was born in 1630 near Rendsburg and was a contemporary of Boyle. He shared with Boyle the process for making phosphorus in 1669 (discovered by Brandt, Kunkel, and Kraft) and for which Hanckwitz (one of Boyle's assistants) & Son established a European monopoly. Kunkel wrote *Laboratorium chymicum* one of the most influential texts of the time and was described by Herman Boerhaave as "the most laborious and renowned chemist, who spent many a tun of gold in making experiments ... to find out the philosopher's stone ... he would have exceeded Boyle: he found out many a curious experiment in chemistry but neglected them on account of the stone".

We must be careful not to view this search for the stone as absurd. Even Joseph Priestley wrote to Benjamin Franklin suggesting that he (Priestley) was about to start a search for the stone. This was, of course, a joke between the two friends with Franklin replying that if Priestley finds the stone he should "take care to loose it again; for I believe in my conscience, that Mankind are wicked enough to continue slaughtering one another as long as they can find Money to pay Butchers" (Benjamin Franklin 1777). If a material with the colour (and there are many natural variations), density, ductility and reactivity of gold could be found then it would not have been possible given the analytic techniques of the day to differentiate it from gold, *i.e.*, there were no mass spectrometers. Even Isaac Newton (1642–1727) spent several months searching for the philosopher's stone and worked with Boyle on the transmutation of mercury. He is reported to have kept his furnaces going day and night and there can be little doubt that he was researching transmutation even while he was Master of the Royal Mint. Newton's will refers to "chymical glasses in the fore garret" and he apparently got hold of Boyle's papers (after Boyle's death) in the belief that the old aristocrat had in fact found the big secret but was wise enough to keep it to himself. By the time of Newton's death, Boyle's philosophy had taken hold in chemistry and the emphasis was turning to analytic science.

A number of exceedingly gifted and talented investigators carried the programme forward. Herman Boerhaave (1668–1738) was Rector at the great Dutch University at Leyden and was renowned for the clarity of his teaching; Peter the Great attended his lectures in 1715. His *Elements of Chemistry* was published in 1732 and was one of the most important texts of the eighteenth century. He made a number of significant advances in medicine and recorded experiments to disprove the claims of alchemists keeping mercury at 100 degrees Celsius for fifteen years with no change. Stephen Hales (1677–1761) devised a method of collecting gases in a bell jar over water in a bucket thereby making a step towards the pneumatic trough. He prepared ammonia from sal ammoniac and lime and experimented in distillation, fusion, combustion and, of course, fermentation. Hales's work was eclectic in its coverage but contained

Stephen Hales (*left*) and Henry Cavendish (*right*)

very little analysis. He did however develop techniques that would prove valuable in the collection of gases and which indeed were still part of the experimental program in the British grammar school chemistry curriculum through the 1960s. The gas collection techniques were further refined and developed by that giant of science, Henry Cavendish (1731–1810). He isolated and characterised Boyle's 'inflammable air' (hydrogen) in 1766, some one hundred years after Boyle's first qualitative observation. He noted that combustion of hydrogen produced a dew-like water. Cavendish also investigated air and found a small volume that would not react chemically. The discovery of the noble gases is rightly attributed to Ramsay but the path was lit by Cavendish.

It was during this period that the quest to understand the driving force behind chemical reactions was initiated. The theory of phlogiston theory played no useful role in the development of chemistry. Professor Woodcock's talk will revisit phlogiston and address Priestley's position on that seriously flawed theory. Suffice it here to mention that the establishment of the Law of Conservation of Matter and ideas on oxidation and reduction in no way addresses the question as to the nature of chemical driving forces.

The importance of an open refereed literature in the explosive development of science during this period can hardly be exaggerated. Boyle's work represented a milestone on the path that led to a true paradigm shift that gave birth to the modern practice of chemistry. He spearheaded the establishment of a peer-reviewed chemical literature that actually reported the results of experiments and provided reliable "histories" for future scientists. One of these future scientists was Joseph Priestley. Priestley was familiar with reports on gases made by Jan van Helmont of Brussels. Van Helmont reported that he had put quite a lot of soda and vinegar in a bottle and observed the formation of bubbles which he named 'gas'. Priestley read Boyle's works carefully and in fact had written from England to Benjamin Franklin requesting Franklin's assistance in obtaining some of Boyle's books which were very expensive.

A number of workers had been successful in generating gases but these always mixed with air and there had been little success in preparing pure gases. For example, Dr. Joseph Black (1728–1799) of the University of Edinburgh had prepared a colourless, odourless gas by heating limestone strongly and this gas when bubbled into clear limewater turned the limewater milky. No other gas that he could find would do that. He acknowledged that he had not collected any of the gas coming from the limestone by heating, but he thought it was the same gas that came from limestone when vinegar was added to it. He called the gas fixed air. Joseph Black was an Ulsterman whose father was in the wine business in Bordeaux. Black was educated in Belfast and apart from his meticulous studies in gases (he discovered carbonic acid and the lime water test for carbon dioxide) also made important contributions in thermodynamics with the discovery of latent heat.

Priestley realized that the previous work on gases had been too ad hoc and lacked any real experimental plan. For example it had not been possible to collect water-soluble gases due to 'suck back' in the pneumatic trough. Priestley thought to use mercury as the liquid collection medium. The non-chemist reader should be warned that the use of mercury in this application has nothing whatsoever to do with the use of mercuric oxide/mercury in the famous experiment in the production of oxygen; a point on which even historians of chemistry have been known to become confused! The plan was successful, and ammonia was collected for the first time. In this way he also obtained for study and experiment the soluble gas that he called "marine acid air," which we know as hydrogen chloride. Then as a special test for this particular gas he said that it gave a white smoke when brought in contact with ammonia gas. (This smoke is used for stage theatricals today, or for the smoke of electric trains.)

Both Boyle and Priestley reported their findings in the Transactions of the Royal Society. Boyle was a founding member (who refused the Presidency because he would not swear an oath) while Priestley holds the distinction of having been expelled. The Royal Society enjoys immense national and international prestige and was beacon in the establishment of a reliable literature and insistence on experimental evidence (the motto is "Nullius in Verba"). The Society was not, however, without critics in the early days. Science was an occupation that was funded from the private purse and could take its toll both socially and financially. The lament of the widow Gimcrack on the passing of her dear husband captures her opinion of the Royal Society,

> When I marries this gentleman he had a very handsome estate, but upon buying a set of microscopes he was chosen a Fellow of the Royal Society, from which time I do not remember ever to have heard him speak as other people did, or talk in a manner that any of his family could understand
>
> Letter from the widow Gimcrack, The Tatler no. 221

Here is Priestley's report in the Transactions of the Royal Society on the production of oxygen:

> Having been given a lens of twelve inches diameter and twenty inches focal length, I proceeded with great alacrity to examine, by the help of it, what kind of air a great variety of substances, natural and artificially prepared, would yield. With this apparatus, after a variety of experiments, on the first day of August 1774, I endeavoured to extract air from mercurius calcinatus. And I presently found, by means of the lens, that air was expelled from it very readily. Having got about three or four times as much as the bulk of my materials, in collecting the air over mercury, I bubbled some of the air into water and found that the air was not dissolved by it. But what surprised me more than I can well express, was that a candle burned in this air with a remarkably vigorous flame, very much like the enlarged flame with which a candle burns when near heated nitre. As I knew no nitre was used in the preparation of mercurius calcinatus, I was utterly at a loss to account for it.
>
> From the greater strength and vivacity of the candle in this pure air, it could be conjectured that the air might be peculiarly salutary to the lungs in certain morbid cases when the common air would not be sufficient to carry off the putrid effluvium fast enough. You will not wonder then that, after having ascertained the superior goodness of the new air by mice living in it and other tests, I should have the curiosity to try it myself. I have gratified that curiosity by breathing it, drawing it through a glass siphon. The feeling of it to my lungs was not sensibly different from that of common air. But I fancied that my breath felt particularly light and easy for some time afterward. Who can tell but that, in time, this pure air may become a fashionable article in luxury? Hitherto only two mice and myself have had the privilege of breathing it.

At around this time, the French chemist Antoine Lavoisier had been reading some old reports in the Transactions of the Royal Society that his attractive young wife, Marie Anne, had translated into French.

Antoine-Laurent and Marie-Anne Lavoisier

They related to the heating of tin in a closed flask as reported by Robert Boyle almost a century earlier. Lavoisier repeated Boyle's experiment except that he selected a larger flask and sealed the flask with the tin in it tightly before applying any heat. He saw that as the heating took place the greyish-white tin oxide (calx) appeared on the surface of the melted tin in precisely the fashion that Boyle had described. He continued the heating for a day and a half, and then allowed the flask to cool just as Boyle had done. However, Lavoisier made two additional measurements that were critical. He weighed the flask before it was heated and he weighed it again after heating, when it was cooled down but before the seal was broken. The weights before and after heating were identical. But when the seal was broken and the oxide crucible was weighted the tin had gained weight. This was the critical point for it meant that whatever was responsible for the gain in weight came from inside the flask and the only thing inside the flask was air. When the flask was opened there was a rush of air into the flask just as Boyle had described. Lavoisier repeated this experiment a number of times even confirming that one-fifth of all the air that had been in the flask was lost. Lavoisier attempted to reduce the tin oxide by heating, but was unable to do so. He also tried the experiment with iron and lead but with negative results. Notice here the persistence in the work and a certain doggedness to prove what he at the time would have suspected to be true. He surely must have had a strong suspicion that the active ingredient was oxygen but was unable to isolate the gas and prove the other half of the experiment. Lavoisier sent a terse report of his experiments and findings (not mentioning Boyle's previous work) to the French Academy of Sciences with instructions that it was to be placed in the files unopened until he asked for its release.

Not long after these experiments were concluded, Joseph Priestley came as a visitor to Paris, and Lavoisier and his wife attended a dinner meeting of scientists at which the English scientist spoke briefly of a discovery, not yet published, that he had made of a strange and wonderful gas formed by the heating of orange-red mercuric oxide. Priestley has noted in his autobiography,

> As I never made the least secret of anything that I observe, I mentioned this experiment also, as well as those with mercurius calcinatus, and the red precipitate, to all of my philosophical acquaintances in Paris, and elsewhere; having no idea at the time, to what these remarkable facts would lead.

Lavoisier realized immediately to what these remarkable facts would lead and promptly repeated his experiments using mercuric oxide thereby producing the first unequivocal evidence for the Law of Conservation of Matter.

Three people discovered oxygen at almost the same time. Two of these were Priestley and Lavoisier. The third was Karl Wilhelm Scheele. Scheele, a largely self-taught chemical genius, died at the early age of forty-four, possible poisoned by mercuric oxide. Lavoisier died under the blade of the guillotine performing his last experiment. After his head was severed he had arranged with an assistant to keep blinking until his severed head lost consciousness; the result was thirteen blinks. Priestley returned to his church duties at Birmingham where his laboratory and home were completely destroyed by a mob whose leaders objected to his support of both the American and French Revolutions. The Priestleys sailed for America. On arrival in Philadelphia, Thomas Jefferson met them and George Washington sent a message of greeting. Joseph Priestley died in 1804, his work on oxygen having played a critical role in the synthesis of Lavoisier and bringing to closure to the programme started by Boyle 150 years previously. After this lengthy gestation chemistry was born.

THE ECLIPSE OF NEWTON: THE DYNAMICAL CHEMISTRY OF JOSEPH PRIESTLEY, 1733–1804

David Knight
University of Durham

David Knight

In the preface to his *History and Present State of Electricity* Priestley ringingly declared his faith that the new sciences in which he was playing a part would prove more fundamental even than Newtonian mechanics had done. There was another scientific revolution underway: and whereas Newton's *Principia Mathematica* [1687] was a forbiddingly recondite publication, electricity, chemistry and optics were accessible. Priestley was a modest man, and his view of science was democratic: but the new science was momentous as well as exciting:

> Hitherto philosophy has been chiefly conversant about the more sensible properties of bodies; electricity, together with chymistry, and the doctrine of light and colours, seems to be giving us an inlet into their internal structure, on which all their sensible properties depend. By pursuing this new light, therefore, the bounds of natural science may be extended, beyond what we can now form an idea of. New worlds may open to our view, and the glory of the great Sir Isaac Newton himself, and all his contemporaries, be eclipsed, by a new set of philosophers [scientists], in quite a new field of speculation. Could that great man revisit the earth, and view the experiments of the present race of electricians, he would be no less amazed than Roger Bacon, or Sir Francis, would have been at his ... What inexpressible pleasure would it give to a modern electrician, were the thing possible, to entertain such a man as Sir Isaac for a few hours with his principal experiments!

Priestley's particular kind of Unitarian Christian materialism meant that the thing was not possible – when we died we ceased to exist, but would by a miracle be resurrected at the Last Judgement. He was no anti-Newtonian. Bacon and Newton had for him laid down the lines along which science should go: and Newton's *Opticks* had been the model for his friend Benjamin Franklin's electrical researches. But there would be more Newtons, as electricity, chemistry and optics became full-blown sciences. And for Priestley and his successors, chemistry was delightful: people flocked to lectures and demonstrations. Although chemists had long tried to improve upon nature, amid debates about the status of artificial and natural products, Priestley's generation met none of the suspicion of the 'chemical' so characteristic of our own day, where the word unfortunately (and rather absurdly) just calls up weapons, pollution and dangerous additives.

Priestley's book on electricity was a great success, translated into French in 1771; but its companion on *Vision, Light and Colours* [1772] was less well received, and Priestley's attention was by then already shifting towards chemistry: especially the chemistry of 'airs' – in 1772 he also published his pamphlet on making soda-water. His significance there was that he brought a new state of matter into chemistry: gases, which emerged not simply as more or less good air, but as chemical elements and compounds as different from each other as solids and liquids. He recognised the part which his vital, eminently respirable, dephlogisticated 'air' (oxygen for short, to Antoine Lavoisier and us) played in life; and the role of plants, given water, in absorbing the 'fixed air' (our carbon dioxide) that animals breathe out, and giving off oxygen. He thus elucidated a crucial chemical cycle involving matter in all three phases. He was no narrow-minded scientist: his friendship with Benjamin Franklin and Thomas Jefferson, as well James Watt and Josiah Wedgwood, his sympathy with the French Revolution of 1789, and his religious views ('rational dissent') made him a celebrity, much involved in controversy.

Priestley's mother died in 1740, and he was brought up by his very religious aunt. As a nonconformist, he could not have gone to Oxford or Cambridge, but got a more modern education at the Dissenting Academy in Daventry, and embarked on a career as a minister. He stammered, and so would never shine as a preacher; but he developed a wonderfully clear and accessible style of writing, and published an enormous amount, mostly religious, political or educational rather than strictly scientific. Professional science did not yet exist; and anyway he saw his role as a minister as central, with science its handmaid, revealing God's plan and activity. He married Mary Wilkinson, daughter of a great ironfounder, in 1762, by which time he was teaching at Warrington Dissenting Academy. Meeting him, Franklin urged him to write on electricity; and his *History and Present State of Electricity* appeared in 1767. By then, he had been elected a Fellow of the Royal Society, and awarded a doctorate by the University of Edinburgh for inventing what he called a chart of biography, and we a time-line.

In 1767 he moved to Mill Hill Chapel in Leeds. There, he wrote the big book on *Vision, Light and Colours*; but it sold disappointingly, and optics was perhaps becoming more mathematical and less open to Priestley's 'hands-on' approach to science. Living next to a brewery, however, he investigated the 'air' that bubbled from the fermenting beer, and began to study gases – inventing soda water, as a means of making artificial mineral water, and thus starting the fizzy drink industry. This was the 'fixed air' that Joseph Black had identified in limestone in 1755: our carbon dioxide. In 1773 the prominent politician Lord Shelburne invited Priestley to be librarian at his great house at Bowood in the summer, and in London in the winter, tutor to his sons, and adviser; with leisure for scientific research. This patronage speedily led in 1774 to the isolation of what we call oxygen. Priestley knew that when mercury was heated in air it formed a powdery red calx, but found that when this was further heated with a burning glass it gave off air in which mice could live, or candles burn, much longer. Priestley, very surprised, gratified his curiosity by 'tasting' it himself, breathing it in and out until he reduced it "to the standard of common air": "I fancied that my breast felt peculiarly light and easy for some time afterwards. Who can tell but that, in time, this pure air may become a fashionable article in luxury. Hitherto only two mice and I have had the privilege of breathing it." Unknown to Priestley, Carl Wilhelm Scheele in Sweden had done the same slightly earlier, but it was not published: such simultaneous discovery is a feature of science, and two eminent chemists have written a play about this one, raising the whole question of what constitutes a 'discovery'. Shelburne in 1774 headed a diplomatic mission to Paris, and took Priestley along. Meeting Lavoisier (who was keen to improve the language and theory of chemistry) and other Academicians, Priestley told them of his discoveries. Back in Britain, he continued his experiments, cheerfully describing his missteps and serendipities as well as his methods of manipulation. He came to realise that gases were not just better and worse air, but were distinct kinds – though, averse to jargon, avoiding the word 'gas'. His time at Calne was very fruitful. Refining his techniques, he collected 'airs' which dissolved in water over mercury instead: in 1790 he reorganised his many publications into a three-volume set of *Experiments and Observations on Different Kinds of Air*.

He loved to emphasise the role of chance in his discoveries. This should not be taken too seriously: it was part of his strategy to make science popular, rather than arcane. Priestley believed, with Francis Bacon long before, that everybody could add their brick to the great edifice of science, and that too many had made a mystery out of it – just as clergy had with religion. Science, especially chemistry, was accessible – and enjoyable. Priestley also shared Bacon's faith that scientific knowledge was power, valuable in industry as his friend Watt was demonstrating; and would strengthen true religion and just government, while freeing everyone from superstition and tyranny. We know that happy accidents only turn into science when the prepared mind encounters them; but there is no doubt that Priestley was propelled by a boyish curiosity rather than by adherence to theory and deduction. Like chemists right through the nineteenth century, he was and needed to be good with his hands. Chemistry was a science of colours, smells and tastes, often indescribable: chemical education was a practical business, and thinking with eyes, nose and fingers was essential. Wedgwood provided ceramic apparatus for Priestley, other friends subscribed to get him glassware and other things he needed, and he clearly enjoyed playing with these toys.

Priestley was for the most part a qualitative worker, though he devised a way of replacing the mice and candles in testing air by oxides of nitrogen, in what was called a eudiometer, giving quantitative results – in principle at least. It was even used on shipboard to test the quality of the air in the bowels of the ship. Lavoisier, a wealthy man, could afford the most expensive apparatus for his laboratory at the Paris Arsenal, and prided himself upon his precision: for him, true science was a matter for experts with superb equipment, an elite activity. He also saw the importance of strictly testable theory in science. Priestley followed Bacon and John Locke in believing that hypotheses had their uses, but should not be taken very seriously. Experiment was a way of finding things out, rather than testing deductions from a theory. The time for a great generalisation had not yet come; and when it did the Newton of chemistry would hit upon some great synthesis involving forces, as Sir Isaac had done, rather than just weights. Priestley like his contemporaries took as his guide the notion that everything that would burn, that most obvious of chemical reactions, contained phlogiston (from the Greek for flammable). When things were set on fire, the phlogiston escaped: either until it had all gone, or until the air in the neighbourhood was saturated with it. Thus metals were compounds of their purified ore ('calx') with phlogiston. For Priestley, his eminently respirable air was dephlogisticated, ready to absorb much phlogiston. Priestley believed to the end of his days that this was the simplest and best explanation, much preferable to Lavoisier's innovative scheme, which he compared to Descartes's imagined vortices of aether carrying the planets around the Sun – an idea happily exploded by Newton.

Lavoisier published his rival account in his *Elements of Chemistry* of 1789, the year the Bastille fell. He called his work a revolution; Priestley welcomed the political but not the chemical revolution. Lavoisier took up Priestley's discoveries, and repeated his experiments; but he saw their significance differently. For him, phlogiston was a looking-glass theory: metals for him were simple substances, he perceived burning as a process in which something from the air was absorbed, and his careful weighings showed that phlogiston if it existed would have negative weight. This to him was absurd; but hydrogen balloons were the latest craze, and to those who did not think very hard about it hydrogen seemed to have negative weight. Some even identified hydrogen with phlogiston. And in naming oxygen (from the Greek, sour-maker), Lavoisier identified it as the cause of acidity. This was understandable. Sulfuric acid and nitric acid are rich in oxygen, and while the lower oxides of metals like chromium and manganese are basic, their higher oxides are acidic. But twenty years on Humphry Davy, the man Priestley saw as his successor, established that the 'marine' or 'muriatic' acid gas that Priestley had isolated using sea salt contained no oxygen; and that the green gas obtainable from it was neither 'dephlogisticated marine acid' nor 'oxymuriatic acid', but an element to which Davy gave the name 'chlorine' on account of its colour. Acidity did not depend on the presence of oxygen. No scientist ever perceives the full implications of a discovery, for science is an ongoing process: but for Davy, as for Priestley, properties were the outcome of forces rather than properties inherent in particular forms of matter. For them, science must be dynamical.

Moreover, Lavoisier invoked 'caloric', the substance of heat, as a weightless element, entering into substances in definite quantities as they changed state, from solid to liquid and from liquid to gas. This seemed to Priestley every bit as mysterious as phlogiston. Indeed, if with the distinguished Victorian chemist William Odling*, we view phlogiston as a groping towards the idea of chemical energy then we can see perhaps why Priestley was reluctant to give it up: anything that burns is indeed rich in chemical potential energy. Priestley,

* A version of this paper was delivered at a conference on Priestley in Leeds on 8th May 2004, and is published in Education in Chemistry, July 2004.

despite his support for the French Revolution and hopes for a more democratic England, was also a staunch patriot; and he disliked the French theory, and the neologisms of Lavoisier and his associates that form the basis of our present chemical language. 'Copper sulfide' and 'lead nitrate', terms which to us seem clear and obvious, were for Priestley awkward terms of art. That means that in reading Priestley's chemistry, we find the need to make an imaginative effort to follow what is almost in a foreign language: what to contemporaries was an admirably plain style has ironically become opaque when applied to the discussion of his beloved chemical reactions. Compared to Lavoisier, Priestley thus seems archaic. The conservatism he showed over theory has overshadowed the fame that the originality, ingenuity and extent of his work on gases brought in his lifetime. The contrast between the coolly reasoning Lavoisier and the impulsive and outgoing Priestley is striking: Priestley making amazing discoveries, and Lavoisier seeing how they entailed a reconstruction of chemistry. But it would be as wrong to see Priestley as a mindless, stick-in-the-mud reactionary as to view him as luckily blundering into fruitful experiments.

In 1779 Shelburne remarried, his boys were grown up, and he pensioned Priestley off. Moreover, Priestley's zealous promulgation of radical politics and religion was proving a bit much for Shelburne: he needed to be quit of this particular sidekick. Moving to Birmingham in 1780, as minister of the New Meeting, Priestley associated with Watt, Wedgwood and others in the Lunar Society, a group that dined together regularly at full moon (so that they could see the way home). These were happy years, in which he published much, especially in politics and religion, but also in chemistry. He connected his science and his religious convictions through his dynamical belief that atoms were centres of forces and matter inherently active. Thus he denied that we were souls imprisoned in bodies, and promulgated materialism – which he believed to be truly Christian. It was in his *Disquisitions relating to Matter and Spirit*, first published in 1777, that he proposed the idea that we do not survive death, but by a miracle we will be resurrected at the Last Trumpet, and judged. We do not go straight to heaven (or hell) for immediate family reunion, as those who believe in immortal souls hope we shall. More significantly for science, he embraced the theory of the Croatian Jesuit Roger Boscovich that atoms were mere points, surrounded by spheres of repulsive and attractive forces. 'Impenetrability' was an illusion: the repulsive forces rose towards infinity as distances got very small, so that no two atoms could occupy the same spot – but chemical affinity, the cohesion of matter, and even thought were all the results of active matter. There was no need for spirit or soul. Boscovich was outraged to see his theory thus taken up by a heretic; and Priestley's ideas found little favour in his own lifetime, as the French Revolution and Reign of Terror were identified as the fruits of materialism. But Priestley's lead was later followed by Michael Faraday, a much more orthodox Dissenter, in the famous 'speculation' in which he put across to a Royal Institution audience his own dynamical ideas about forces and fields. Later, Clerk Maxwell and William

Thomson (Lord Kelvin) also took this kind of matter-theory very seriously: it did duly transform the Newtonian world of impenetrable particles and central forces into a new physics in due course.

Priestley never presented himself as 'the Newton of Chemistry', but in the next generation Davy and Dalton perhaps did so. Davy in his researches on laughing gas and the other oxides of nitrogen began in Priestley's tradition, under the aegis of Priestley's Lunar friends Wedgwood and Watt, with whose sons Gregory and Tom he became friends. This was 'suck it and see' chemistry, where breathing nitrous oxide led to wonderful sensations beautifully described: Davy waved his arms, laughed wildly, and proclaimed that nothing exists but thoughts. Breathing carbon monoxide had different effects: when brought round by his friends, Davy weakly declared that he did not think he would die. Davy carefully analysed the various oxides of nitrogen, and his book describing the experiments established his reputation, and brought him an enthusiastic acknowledgement from Priestley, whose mantle thus fell upon him. Davy had written 'Davy and Newton' in a youthful notebook; and Alessandro Volta's work enabled him to realise Priestley's vision, with the idea that chemical combination is electrical. His famous experiments on the electrolysis of water, establishing that hydrogen and oxygen are the only products under properly controlled conditions, and isolating potassium from potash, were goal-directed researches in the spirit of Lavoisier rather than Priestley. The same was true of his celebrated work on flame leading to the miners' safety lamp: a pioneering example of 'applied science' going beyond happy accidents or hands-on inventiveness. These researches led to Davy sitting in Newton's chair as President of the Royal Society, the country boy now at the centre of things; a very different fate from that of Priestley, though it could not be said that Davy's tenure was particularly happy.

Dalton's training had been mathematical, and he was a great admirer of Newton – his spare copy of the *Principia Mathematica* [1687], passed on to a friend, is now in Durham University library. His vision of science was hypothetico-deductive, whereas Priestley's was inductive, involving generalisations and working hypotheses rather than structured theory. Around 1800, everyone believed in the corpuscular atoms of Robert Boyle and Newton: minute particles of matter, of one or very few different kinds, which in various arrangements would compose the various metals and other elements. Nothing much followed from this commitment. For Dalton it was different: each element for him had irreducibly-different chemical atoms. So, where Davy had analysed the oxides of nitrogen and expressed the results as percentages, indicating nothing but recipes; Dalton with his particular atomic theory thought of recalculating them as the amount of oxygen combined with a fixed quantity of nitrogen, and found simple ratios. His theory enabled him to make sense of empirical data. Like Priestley, Dalton was a dissenter, uneasy in the polished world of London society: electricity, which had galvanised Davy, was less his interest than optics, where he famously investigated his own colour-blindness.

These two maintained the momentum of Priestley's work, and established that chemistry was not exclusively a French science. For Priestley meanwhile things had not worked out happily. Having welcomed the French Revolution, on 14th July 1791 Priestley was wrongly believed to be at a banquet celebrating the second anniversary of the fall of the Bastille; and a mob shouting "Church and King" rampaged through the city and sacked his house. He fled to London, to Hackney where there was a Dissenting Academy at which he taught; but there too sympathy with France had evaporated. Although he got some compensation for his losses, Fellows of the Royal Society kept their distance from the heretic and firebrand, denounced by Edmund Burke in Parliament as keen to blow up the establishment with his new chemistry, and pilloried by cartoonists. Lavoisier, in contrast to Priestley, was extremely rich. His family was well-off, he married an heiress, and he played an important and lucrative part as a 'tax-farmer' in the privatised system of pre-revolutionary France. In May 1794 Lavoisier and the other tax-farmers were guillotined as fat-cats. The story that the judge remarked that the state had no need of scientists seems a fiction, because other chemists unsullied by their past prospered, and Paris remained the world's centre of excellence in the sciences throughout the years of revolution and war lasting until 1815. Lavoisier was ungenerous to his contemporaries, French and foreign, never really admitting how much he owed to Priestley and others. Thus he misleadingly wrote of oxygen: "This species of air was discovered almost at the same time by Mr Priestley, Mr Scheele, and myself". But his sharp intellect, grasp of the need for theory, and leadership had transformed chemistry by the time he died.

Also in the spring of 1794 the Priestleys decided to escape hostility and join other family members in the USA. Thus Lavoisier and Priestley, the leading chemists of the day, were both rejected by their societies, for political reasons; and Priestley was the first in a long line of distinguished refugee scientists making a new life in America. The Priestleys decided to live not in the busy and flourishing city of Philadelphia, but in the small town of Northumberland up the Susquehanna River, where the young romantic poets, (and associates of Davy) Samuel Taylor Coleridge and Robert Southey contemplated joining him. But Mary Priestley died, and Joseph felt an exile, homesick and never taking US citizenship. An enthusiastic, modest and sociable man who had made some of the greatest discoveries in the history of chemistry passed his last ten years in relative isolation in a strange land. He never changed his mind about Lavoisier's chemistry, publishing in 1800 a book, *The Doctrine of Phlogiston Established*, which put him out of step with the majority of chemists in Britain by then. He went on experimenting, and publishing papers in the USA and Britain, to the end of his life; but by then he was out of the mainstream.

As well as oxygen, Priestley first isolated numerous other gases, using the technique of collecting over mercury those that dissolved in water. Thus among his discoveries were the now well-known gases we call hydrochloric acid (1772), nitric oxide (1772), ammonia (1773), sulfur dioxide (1774), and carbon monoxide (1784). He described their properties in the classic three-volume set of *Experiments and Observations on Air*, published in Birmingham in 1790, not long before the catastrophe. The book, based upon his papers published over the years, is still informal, even exuberant, and he reports for example "thinking of this most extraordinary fact upon my pillow" as well as giving experimental details, and reasoning upon them. Scientific monographs are not written that way now. His vision of chemistry as essentially experimental, open to everyone, and fun made it the most attractive science of the early nineteenth century, when Davy for example drew great crowds to his lectures in Regency London. Lavoisier had sought to make chemistry more like algebra, and was an admirer of experimental physics. The years around 1800 were a period of scientific revolution, as careers opened up bringing professional science and formal training, mathematics became more prominent, and Priestley's life as a minister and politician, and his cheerfully experimental science, came to seem amateurish. He never got a law or a unit named after him, like Robert Boyle, Amedeo Avogadro, or Alessandro Volta; and so he has not always been placed, as he ought to be, not only among the most famous men of his time but also among the greatest (as well as one of the most attractive) of chemists.

References

J. Priestley, *The History and Present State of Electricity, with original experiments*, 3rd ed 1775, reprinted New York: Johnson, 1966, 1, xv.

I. B. Cohen ed., *Benjamin Franklin's Experiments*, Cambridge MA: Harvard U.P., 1941.

W. R. Newman, *Promethean Ambitions: Alchemy and the Quest to Perfect Nature*, Chicago: Chicago U.P., 2004.

J. Priestley, *The History and Present State of Discoveries relating to Vision, Light, and Colours*, London: Johnson, 1772.

P. Wood ed., *Science and Dissent in England, 1688–1945*, Aldershot: Ashgate, 2004.

R . E. Schofield (ed), *A Scientific Autobiography of Joseph Priestley (1733–1804): Selected Scientific Correspondence*, Cambridge MA: MIT Press, 1966.

D. C. Lindberg and R. Numbers ed., *When Science and Christianity Meet*, Chicago: Chicago U.P., 2003; J. H. Brooke and G. Cantor, *Reconstructing Nature: the Engagement of Science and Religion*, Edinburgh: T. & T. Clark, 1998.

J. Priestley, *Experiments and Observations on Different Kinds of Air, and other branches of natural philosophy connected with the subject*, [1790], reprinted New York: Kraus, 1970, 2, 162.

C. Djerassi and R. Hoffmann, *Oxygen*, Weinheim: Wiley VCH, 2001.

A. David, F. Fernandez-Armesto, C. Novi and G. Williams ed., *The Malaspina Expedition, 1789–1794*, Vol. 2, London: Hakluyt Society, 2003, pp. 249, 253.

J. Golinski, *Science as Public Culture: Chemistry and Enlightenment in Britain, 1760–1820*, Cambridge: C. U.P., 1992.

R. G. W. Anderson and C. Lawrence ed, *Science, Medicine and Dissent: Joseph Priestley (1733–1804)*, London: Science Museum, 1987.

J. Priestley, *Heads of Lectures on ... Chemistry, delivered at the New College in Hackney*, London: Johnson, 1794, p. 3.

H. Davy, 'Researches on the Oxymuriatic Acid', *Philosophical Transactions*, 100 (1810), 231–57; 'The Bakerian Lecture', *Philosophical Transactions*, 101 (1811), 1–35; esp. 32.

J. Uglow, *The Lunar Men: the Friends who made the Future*, London: Faber and Faber, 2002.

D. M. Knight, *Science and Spirituality: the Volatile Connection*, London: Routledge, 2004, 8–19.

J. Priestley, *Disquisitions relating to Matter and Spirit*, 2nd ed., Birmingham: Johnson, 1782.

M. Faraday, 'A Speculation touching Electric Conduction and the Nature of Matter', *Experimental Researches in Electricity*, vol. 2, London: Taylor, 1844, pp. 284–93.

J. C. Maxwell, *Scientific Papers*, ed. W. D. Niven, Cambridge: CUP, 1890, vol. 2, pp. 448, 471, 480; B. Mahon, *The Man who Changed Everything*, Chichester: Wiley, 2003, pp. 122–7, 134.

W. Thomson, *Baltimore Lectures on Molecular Dynamics and the Wave Theory of Light*, Cambridge: C. U.P., 1904, pp. 123, 125, 131, 285, 556, 653, 668, 675. C. Smith and M. N. Wise, *Energy and Empire: a Biographical Study of Lord Kelvin*, Cambridge: C. U.P., 1989, pp. 156, 339, 380.

D. M. Knight, *Humphry Davy: Science and Power*, 2nd ed., Cambridge: C. U.P., 1996, pp. 26–41.

J. Z. Fullmer, *Sir Humphry Davy's Published Works*, Cambridge MA: Harvard U.P., 1969, p. 29.

A. L Lavoisier, *Elements of Chemistry*, [1790], tr. R. Kerr, reprinted New York: Dover, intr. D. McKie, 1965, p. 36.

CHEMISTRY AND REVOLUTION

Colin Russell
Open University and University of Cambridge

Introduction

Colin Russell

Thomas Carlyle once wrote: "All things are in revolution; in change from moment to moment, which becomes sensible from epoch to epoch: in this Time-World of ours there is properly nothing else but revolution and mutation, and even nothing else conceivable".

Modern (*i.e.* Lavoisierian) chemistry was born in a period of great political upheaval and revolution. The American Revolution had culminated in the Declaration of Independence in 1774. Much nearer to home the French Revolution had begun with the fall of the Bastille in 1789. The fervour and mounting hysteria led to that darkest period in French history, the Terror of 1793–4. Multitudes were killed by the guillotine, and Louis XVI was beheaded in January 1793.

In Britain, these events over the Channel caused strong reaction. Revulsion to the Terror led to declaration of war with France in 1793. At once the Home Office became concerned with preventing sedition, and a Royal Proclamation was soon issued against seditious writings. In 1794 trials for treason and related offences began. Nor were matters improved in 1795 when George III was attacked by a mob, though managed to escape unscathed.

Effects of political revolutions on chemistry

1. At a personal level

(a) Lavoisier

Lavoisier was condemned by revolutionaries for being a Tax Farmer. The question was not so much whether he had been fraudulently involved (there is no evidence that he had), but rather the nature of his job. With his father-in-law and several others, he was under arrest by Nov. 1793. In prison he displayed great dignity and calm. He wrote to his wife "Yesterday you appeared to be sad; why should you be, when I accept what will happen with resignation ... there is also reason to hope we shall be together again".[1] Next morning, 4th May, 1794, he went to the guillotine – the fourth that day, and just after his father-in-law. Within seconds, the flashing blade had shorn away the life of France's greatest chemist.

A rumour that his enemies jeered at him with the words "The state has no need of savants" may be apocryphal, but their intention was clear enough.

Lavoisier's legacy was thus a nameless grave, a grieving widow, and a reputation as a hero of science.[2] But chiefly he left behind a chemistry freed for ever from the myth of phlogiston.

(b) Joseph Priestley

Priestley was a dissenting minister, *i.e.* one who did not acknowledge the authority of the established Church of England.[3] This applied both to church government and also (in some cases) to church doctrine. As a Unitarian he did not subscribe to the doctrine of the Trinity, that God was three Persons (Father, Son and Holy Spirit) in one Godhead (though this belief was and is held by many Christians who do not call themselves Anglicans).

Priestley had a knack of making himself unpopular; being a Dissenter may have hardened him to criticism and unpopularity. Most notably for our purposes must be the fact that he supported the French Revolution.

At this time he was living at Fair Hill, Birmingham. In July 1791, a luncheon had been planned in Birmingham to celebrate the Revolution across the Channel. It took place on the afternoon of 14th July (Bastille Day). The enthusiasts drunk no less than 19 toasts but all went peaceably and they had dispersed by 6 o'clock.

Two hours later another crowd assembled to express opposition to the republican supporters. They attacked a tavern, then the chapel of the so-called New Meeting, then that of the Old Meeting, setting some of the property ablaze. It was a warm, moonlit night. From Priestley's house at Fair Hill, about a mile away, the flames were ominously visible. After midnight the mob had reached his house, destroying much of it, including his laboratory. Fortunately the Priestleys had been smuggled out in time.

With their war cry "Church and King", the rioters not only attacked their political enemies but also (as in France) dealt a severe blow to science. Their motivation has been much questioned ever since. In 1793 Wordsworth wrote to Bishop Watson, "Left to the quiet exercise of their own judgment, do you think that the people would have thought it necessary to set fire to the house of the philosophic Priestley?".[4] This has become a standard view of events. It is possible that their feelings were inflamed by a handbill suggesting a "war of freedom" that would ultimately destroy even the monarchy; and this might have been the work of an *agent provocateur*. But evidence that this was a Tory-inspired mob, is otherwise lacking. It looks much more like a case of spontaneous combustion, and in its early stages the protest was even sanctioned by three local magistrates.[5]

As for Priestley himself, he had attacked the establishment of the Church of England (1790) and was known to support the French Revolution. He could hardly have hoped to escape the fury of a crowd who saw the Revolutionaries much as we view modern terrorists and recognised in him an ardent supporter of that cause.

After a brief stay in Hackney the Priestleys took ship for America in 1794, settling at Northumberland, PA, where Joseph died in February 1804. Even there he did not escape political suspicion, and wisely kept his head down at first. He did some experiments on the Voltaic pile and a few other topics but was by now clearly out of touch with much new work. The most astonishing thing about him, however, was that he who had isolated the gas we now call oxygen to the end of his days remained committed to phlogiston. To isolate a substance without comprehending what it was can hardly constitute a full 'discovery' of that substance, which is why Lavoisier much more than Priestley must be credited as the discoverer of oxygen. But historians of science are rightly cautious about use of the term 'discover' because if its inherent ambiguity in certain contexts. It is far better to recognise that Priestley, Scheele and Lavoisier each made valuable contributions to oxygen chemistry, but to leave the title of 'discoverer' out of the discussion.

(c) Beddoes

Thomas Beddoes (1760–1808) was a successful teacher of chemistry at Oxford.[6] He was also an ardent republican, publishing a tract in 1792, that condemned the British Government for issuing "fabricated lies" about the French revolutionary Government, the French clergy for violence and corruption (including treatment of refugees), and the Birmingham mob for wanting to "roast Dr Priestley alive" ...

He had tried (with approval from the Vice Chancellor) to gain a Readership. However, nothing came of it, for he was evidently a marked man.[7] The reason may be found in a letter from Whitehall (November 1792) to the local MP, noting that "Dr. B. has lately been very active in sowing sedition in your neighbourhood, particularly by the distribution of pamphlets of a very mischievous and inflammatory tendency".[8]

Frustrated in his academic ambitions, Beddoes left Oxford in April 1793 for Bristol. Later, he founded his Pneumatic Institution,[9] where Davy cut his chemical teeth and where the anaesthetic properties of nitrous oxide were discovered. But thereafter neither Beddoes nor his Institution contributed much of permanent value to science.

However, he kept on trying to promote his political views. Thus he wrote a letter to William Pitt suggesting ways of alleviating poverty caused by chronic food shortages. Pitt did not reply. Undeterred by the lack of response, Beddoes went on to write an "Essay on the public merits of Mr Pitt". Evidently there were not many!

Beddoes lacked tact as well as self-discipline and his science, like that of Lavoisier and Priestley, fell victim to the political revolutions through which he lived. However these upheavals affected nations as well as individuals.

2. At a national level

One illustration must suffice. After Lavoisier's death his theory of oxygen gained a very mixed reception. Chemical education in France made use of it at an early stage.[10] On the other hand Germans, in particular, disliked oxygen and preferred phlogiston. F. A. C. Gren at Halle was the first to come out against the new ideas, but J. F. Westrumb in Hamelin led the opposition. The age-old hostility between France and Germany affected even chemistry. It has been well said that "nationalistic antipathy to French culture certainly conditioned the initial response of the German chemical community to Lavoisier's system".[11]

Chemistry and revolution are frequently intertwined. This is even more true at an international level.

3. At an international level

Again, only one illustration must suffice. In 1808 John Dalton published his *New System of Chemical Philosophy* where the chemical atomic theory was described. But that was in England. In France, despite the war, one chemist (Berthollet) had a copy within 8 weeks![12] It was fairly easy to cross a mere few dozen miles of water with one small item.

This, however, was not typical. Napoleon had instituted his "Continental system" as a hangover of the revolutionary wars. It was a blockade of ports in neighbouring countries through which assistance might come to the British enemy. As one historian said many years ago, speaking of another age-old rivalry, that between France and England:

> Could not the ruler of Europe exclude British ships from all the harbours of Europe, and would not this reduce Great Britain to starvation? There was something in this policy that was traditionally French.[13]

One country to be so affected was Sweden, and here Berzelius had to wait four years before he could lay hands on a copy of the *New System*. Here and elsewhere, revolution delayed the extension of atomic theory for that length of time. When Davy's famous dialogue with Berzelius eventually took place, it was only after 1812 when Napoleon's power was seriously diminished. That led to the science of electrochemistry, delayed and even shaped by the revolutions in which it was conceived.

It is, of course, a simple mistake to suppose that only political systems were in a state of revolution around 1800. There was also an Industrial Revolution, an Agrarian Revolution, and above all a Chemical Revolution.

The Chemical Revolution

The "Chemical Revolution" is a phrase that is always ascribed to the work initiated by Lavoisier (much more than by Scheele or Priestley). It was the recognition of the role of oxygen in combustion, though first in acids. It was regarded as a French concept early on,[14] and became a cult reflecting political interests and national pride.[15] It could also have been applied to the work of Dalton and atomism. This became a British and Swedish emphasis in the 19th century though the Germans soon adopted it. Even in the 1900s certain schools of chemistry in France uniquely regarded atomism with scepticism.

There is no evidence that formulation of this theory owed anything to revolutionary politics. Rather it sprang from Newtonian dynamics and a Quaker simplicity.

However, all this time something else was happening in chemistry that was arguably at least as important as the revolution of Lavoisier. Sometimes called "The OTHER chemical revolution" it was described in a seminal work of Archie and Nan Clow.[16] It was the extension of chemistry to industrial scale and its application to industrial practice.

The first big breakthrough came with the large-scale manufacture of sulfuric acid in the 18th century. This was followed by the work of the 9th Earl of Dundonald (1749–1831) leading to gas manufacture, eventually on a very large scale. These two processes were at the vary basis of the *Industrial Revolution* that was now under way.

There is another case, rather less well known. Richard Watson (1737–1816), 5th Professor of Chemistry at Cambridge from 1764 became Bishop of Llandaff in 1782. He lived near neither places, but in Westmoreland, at Calgarth Park. Here he did experiments and wrote books. In 1787 the Duke of Richmond (Head of the Government's Ordnance Department) approached him for advice as to possible improvements in gunpowder needed in sea-battles with the French and inferior to theirs in power. He identified the problem as residing in the method of preparing the charcoal.

Traditionally charcoal had been made by heating piles of wood with air excluded as far as possible. Now, however, Watson advised distilling the wood in closed vessels. It was tested at Hythe and "the improvement has exceeded my utmost expectation". A 68lb ball was propelled 273 feet by powder made from the new charcoal, while under identical conditions with 'traditional' charcoal, the distance was a mere 172 feet. Further experiments confirmed that the cylinder powder effected an improvement of 100: 60.[17] Later work showed even greater improvement.

This was obviously due to a variety of causes:

- Temperature: the higher the temperature the higher the proportion of carbon, and also the H/O ratio. But if it gets too hot, the carbon is too dense (small surface area). In addition, charcoal made at too low a temperature tends to be hygroscopic.
- Time: traditional methods may take up to a week; retorts as little as 8 hours.[18]
- Absence of air: clearly less charcoal will be burned.
- Impurities: varying quantities of earthy matter are always present in 'pit' method.

So control was essential. The Government set up cylinder charcoal works in Sussex (two) and in some of the Cumbrian gunpowder works, first at Sedgwick on the R. Kent[19] and at Lowwood early in the nineteenth century.[20] At Waltham Abbey[21] the use of cylinders was early recommended by Major William Congreve but not introduced till 1794.[22]

So what was the effect of these recommendations? An official told Watson that he had saved the Government £100 000 p. a. Presumably this was because there was less wastage in charcoal manufacture, and also because a smaller charge of gunpowder could often suffice.

However, it is likely that there were other less quantifiable results. Did it give an edge of superiority in sea battles? In the Battle of the Nile (1st August 1798), nearly all French ships were destroyed, including the flagship *Orient* and this was when the improved powder was introduced. Obviously, there were many other factors such as nautical skills, weather conditions, morale of sailors and much else. The Royal Navy used carronades, short guns introduced in 1779 and known by the French as "devil guns". They were worked until about 1865 and could fire up to 64 lb. It is hard to imagine that the improved gunpowder was not of significant effect with these weapons. However, much more research is needed before we can be sure.

When the Sedgwick works closed in 1850 six charcoal cylinders were moved to nearby Gatebeck. To this day two of them survive – as *gateposts* to a caravan park![23]

If these speculations are correct the wheel has turned full circle. From a chemistry affected and impoverished by revolution it has now become an agent in the direct conduct of revolutionary wars. And this was possible because it had itself been transformed at the theoretical and the technical level by two changes of such magnitude that we can only call them 'chemical revolutions'.

NOTES

1. A. Donovan, *Antoine Lavoisier*, Cambridge University Press, 1996, p. 299.
2. W. A. Smeaton, "M. and Mme. Lavoisier in 1789: the chemical revolution and the French Revolution", *Ambix*, 1989, 36, 1–4.
3. M. Fitzpatrick, "Science and society in the Enlightenment", *Enlightenment & Dissent*, 1985 (4), 83–105.
4. B. Willey, *The eighteenth-century background*, Pelican, 1940, pp. 247–9
5. C. Emsley, *Britain and the French Revolution*, Longman, 2000, pp. 40–1.
6. T. H. Levere, "Thomas Beddoes at Oxford: radical politics and the Regius Chair in Chemistry", *Ambix, 1981*, **28**, 61–69.
7. T. H. Levere, "Dr. Thomas Beddoes (1750–1808): science and medicine in politics and society", *Brit. J. Hist. Sci.* 1984, **17**, 187–204.
8. D. A. Stansfield, *Thomas Beddoes M. D.*, 1760–1808, Reidel, Dordrecht, 1984, p. 78.
9. T. H. Levere, "Dr. Thomas Beddoes and the establishment of his Pneumatic Institution: a tale of three presidents", *Notes & Rec. Roy. Soc.*, 1977, **32,** 41–44.
10. W. A. Smeaton, "Chemical education in Revolutionary France", *J Roy. Inst. Chem.*, 1958, **82**, 650–656.
11. K. Hufbauer, *The formation of the German chemical community* (1720–1795), University of California Press, Berkeley, 1982, p. 103.

12. M. P. Crosland, "The first reception of Dalton's atomic theory in France", in D. S. L. Cardwell (ed), *John Dalton & the progress of science*, Manchester University Press, Manchester, 1968, p. 277.

13. A. J. Grant and H. Temperley, *Europe in the nineteenth and twentieth centuries*, Longman, 1939, p. 136.

14. J. B. Gough, "Some early references to revolutions in chemistry", *Ambix*, 1982, **29**, 106–109.

15. B. Bensaude-Vincent, "Une mythologie révolutionarie dans la chimie Française", *Ann. Sci.*, 1983, **40**, 189–196.

16. A. and N. Clow, *The Chemical Revolution*, Batchworth, 1952.

17. R. Watson, *Anecdotes*, 1817, I, 241–1.

18. At Faversham 580 lb. of wood in each of 3 cast-iron cylinders was heated for 8 hours.

19. E. M. Patterson, *Blackpowder manufacture in Cumbria*, The Faversham Society, 2nd Impression, 1995, p. 9.

20. G. Cocker, *The Lowwood Gunpowder Works*, R. E. Harvey, Cartmel, 1988, 6.

21. B. J. Buchanan, "Waltham Abbey Royal Gunpowder Mills: 'The Old Establishment'", *Trans. Newcomen Soc.*, 1998–9, **70**, 221–250.

22. E. Gray, H. Marsh and M. McLaren, "A short history of gunpowder and the role of charcoal in its manufacture", *J. Materials Science*, 1982, **17**, 3385–3400.

23. A picture of the gateposts can be found in M. Archer and C. Haley (eds). *The 1702 Chair of Chemistry at Cambridge: Transformation and Change*, C. U.P., Cambridge, 2005, fig. 3.4.

CALORIC, PHLOGISTON AND CHEMICAL POTENTIAL

Les Woodcock
University of Manchester

Introduction

Les Woodcock

Historians of science have not been kind to Joseph Priestley. This presentation will try to re-evaluate one of Priestley's contributions to physical chemistry 200 years after his demise.

Every chemist knows that Priestley discovered oxygen, and some chemists know he also discovered and characterised nine other gases, including ammonia, which he decomposed into hydrogen and nitrogen by electrolysis 75 years before Faraday. His many contributions to physics, however, such as Coulomb's law 20 years before Coulomb, description of atoms 40 years before Dalton, hypothesis of division of atoms 150 years before Rutherford, forces between atoms 100 years before van der Waals, prediction of Black Holes 200 years before Stephen Hawking, are largely unknown. In biology, Priestley also first discovered and described the process of photosynthesis and the carbon cycle. He also discovered the solubility of carbon dioxide in water and went on to invent the process of carbonation or fizzy drinks industry.

Priestley achieved many great firsts in science, but he remains famous only for his discovery of oxygen. In this presentation, however, I am going to concentrate on that aspect of his work for which he became "infamous". Priestley advanced the theory of phlogiston to explain why materials reacted with oxygen, in the way they did. He continually had to refine the theory, but in the end, he died ignominiously in the eyes of the developing community of chemical scientists in the early 19th century ...

Despite this life of scientific discoveries, Priestley died in 1804, "stubborn and stupid", the last of the Phlogistonists. He might have been stubborn, but was definitely not stupid. Here we present a modern interpretation of phlogiston as 'the negative of the chemical potential of a material with respect to its oxidation', and show that phlogiston was the first reasonable scientific theory of chemical change, 100 years before Gibbs got it right. We will also argue that the Antiphlogistonists, by contrast, may have been the founders of 'inorganic chemistry' but contributed little to physical chemistry by failing to address the fundamental question of why chemical reactions occur.

Background to phlogiston

The concept of phlogiston was introduced around 1700 by the German natural philosopher Georg Stahl. Until then it was believed by the alchemists that everything was made of just four elements, earth, water, air and fire. Observations of combustion by various materials however showed that the many different kinds of earth required a fifth element that explained why, for example, some materials burn in air much more readily than others. Different kinds of earth, *i.e.*, solids, were deemed to contain variable quantities of 'phlogiston' which, from Greek, literally means "fire of the earth". The precise definition of phlogiston was not clear; some philosophers regarded phlogiston and fire as being synonymous; when a solid burned, it simply transferred its phlogiston to the air. Phlogiston was a concept that Joseph Priestley adopted, developed further, and adhered to throughout his life as a natural scientist.

Caloric: as a state function

We now know that the understanding of chemical reactions is inextricable dependent upon an understanding of the concept of 'heat'. A big step forward in the 18th century was the introduction of the concept of a state function caloric by Joseph Black. Unfortunately, at that time, scientists did not know the difference between energy, the state function, and heat, which is not internal energy within a material, but energy on the move.

Prior to the advent of thermodynamics around 1850, all scholars of science believed in the caloric theory that heat was a conserved fluid with no mass. It was present in various amounts in all materials and flowed from high to low concentrations. The caloric content depended on temperature and physical state; gases had a high caloric content and solids a low caloric content. The basic misconception was that caloric was a conserved substance *i.e.* it could be "neither created nor destroyed". We now know that what was thought to be 'caloric' is an internal energy. Following the discovery of the First Law of Thermodynamics, the conservation of the sum of heat and work, caloric can now be identified with the state function 'enthalpy'. This concept is developed below.

First chemical equations

By the mid 1700s natural philosophers such as Priestley were beginning to write down chemical equations. Almost all of the earliest chemical reactions that were studied by Priestley and his contemporaries involved oxygen.

The first attempts at chemical equations, as seen in Table 1, generally involved a material containing its phlogiston reacting with air and it giving up its phlogiston for it to appear in one of the products. Besides his most famous discovery of all, oxygen, Priestley went on to discover and characterise the reactions involving nine gases. He isolated pure oxygen by heating the oxide of mercury. In keeping with the phlogiston concept, air that was capable of reacting with a solid to accept the transfer of phlogiston was aptly named as dephlogisticated air.

Table 1: A list of some early chemical equations; the symbol Π *is used to represent phlogiston.*

charcoal (+ Π) + air = fixed air (CO_2) + phlogisticated air
metal (+ Π) + air = calx (pure base) + phlogisticated air
charcoal (+ Π) + calx = metal (+ P) + fixed air
metal (+ Π) + water = calx + inflammable air (Π)
metal (+ Π) + acid = salt + inflammable air (Π)
mercury (+ Π) + dephlogisticated air = calx of mercury
inflammable air (+ Π) + dephlogisticated air = water

Soon after the time of his discovery of dephlogisticated air, Priestley met Lavoisier in Paris and told him of his discovery. Lavoisier, however, had invented a balance that could weigh very accurately to 0.0005 g. He was able to confirm that, when inflammable air reacted with dephlogisticated air to give water, there was no mass present except that of the reactants and products of the reaction. Lavoisier had discovered the 'Principle of Conservation of Mass' in chemical reactions. He argued that since there was no mass that could account for phlogiston, therefore it did not exist. *Phlogiston died on September 5 1775: the day Lavoisier presented his paper to the French Academy of Science.*

Figure 1. Antoine Lavoisier (1743–1794) usurped Priestley's discovery and renamed dephlogisticated air as oxygen

Following the Lavoisier-Priestley meeting in 1774, and as the years went by, more experiments were performed by Lavoisier and his followers and they became known as the antiphlogistonists. With every new experiment, the antiphlogistonists would come up with another objection to phlogiston theory. The phlogistonists, led by Priestley, would modify the theory to fit the new experiment.

Eventually, one by one, all the phlogistonists became antiphlogistonists until, by the mid 1790s, only Priestley was left. He adhered to the idea that chemical reactions were driven by phlogiston until the day he died in 1804; he was widely portrayed, especially by the French antiphlogistonists, and also by his many political adversaries, as being stupid and stubborn.

Figure 2. "Priestley est stupide" was the cry from the anti-phlogistonists. This gave sustenence to Priestley's many political enemies, not least King George III and sycophantic institutions of the monarch such as the Royal Society of London

Gibbs (free) energy

Why do chemical react? In order to understand chemical reactions, we have to move on to the advent of chemical thermodynamics. The First Law of Chemical Thermodynamics was discovered in 1850 by James Prescott Joule; the Second Law and the concept of entropy was proposed around the same time by Rudolf Clausius. It was not until Gibbs introduced the concepts of Gibbs energy G and chemical potential μ that the true driving force for all physical and chemical change was discovered.

In 1885, the American mathematician and engineer Josiah Willard Gibbs finally applied the recently discovered and formulated laws of thermodynamics to explain why one chemical will react with another to form different compounds. Chemical reactions will take place in the direction of equilibrium until the composite Gibbs energy of the reactants and the products is a minimum. Then the reaction ceases.

Ellingham diagrams

A modern platform for explaining the reactions of various elements with oxygen is the Ellingham diagram. These diagrams plot (for a set of elements X) the molar Gibbs energy change ΔG_m ($= \Delta G/\Delta\xi$) for the reaction

$$xX + O_2(gas) \rightarrow y(\text{oxide of } X);\ \text{pressure} = 1\ \text{bar} \qquad (1)$$

against temperature T

The more negative is ΔG_m at any chosen temperature, the greater the propensity of the element to react with oxygen.

For this general reaction, ξ in the graph of Figure 3 is defined as the reduction in the chemical amount (n) of O_2 (a quantity commonly expressed in moles) as the reaction proceeds, *i.e.*, $\Delta\xi = -\Delta n(O_2)$ or equivalently $-\Delta n(X)/x$. Furthermore $\Delta n(X)$ can also be expressed in terms of the change in mass $\Delta w(X)$ of X by

$\Delta n(X) = \Delta w(X)/M$; M = molar mass of X (= molecular weight × g mol^{-1})

Thus

$$\Delta G/\Delta w(X) = \Delta G_m/(-xM)$$

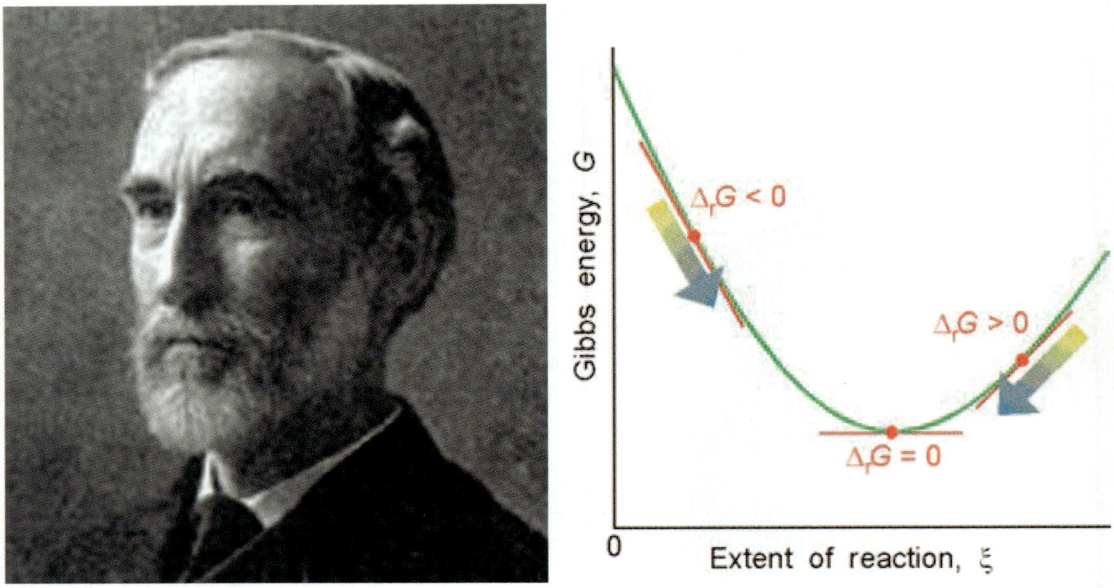

Figure 3. The great American scientist J.W.Gibbs was the first to explain why chemical reactions take place; he introduced the concept of Gibbs (free) energy and the related concept of chemical potential.

The graph shows the natural trend of a chemical process (at constant temperature and pressure) to minimum Gibbs energy.

The more positive this *modified* ΔG_m the greater the propensity of equal masses of X to react with oxygen

Phlogiston: a thermodynamic state function

In order to see the connection between phlogiston and energy, we must now explain the concept of a state function. Energy in the form of heat (denoted by q) is not a property of a material, *i.e.*, a state function; it is energy on the move through or between materials. The total energy content of a chemical is a state function, because it depends only upon its equilibrium state of temperature and pressure. An alternative simple way of stating the First and Second Laws of Thermodynamics is to define the status of the energy as being available or unavailable if burned, in a material that changes when it is heated (or cooled).

First Law: Enthalpy ('heat content') H is a state function $\Delta H = q$
Second Law: Entropy S ('unavailable heat content"/ T) is a state function

$$\Delta S = \tilde{q}/T$$

($\tilde{q}$ is the heat absorbed *reversibly* at absolute temperature T)

Some of the energy content of a chemical may be unavailable as heat when it burns, the amount of unavailable energy depending on temperature at which it exists.

A combination of the First and Second Laws leads to the definition of the Gibbs energy state function which predicts the position of equilibrium of all chemical reactions (*i.e.*, when $\Delta G/\Delta\xi = \Delta G_m = 0$) then

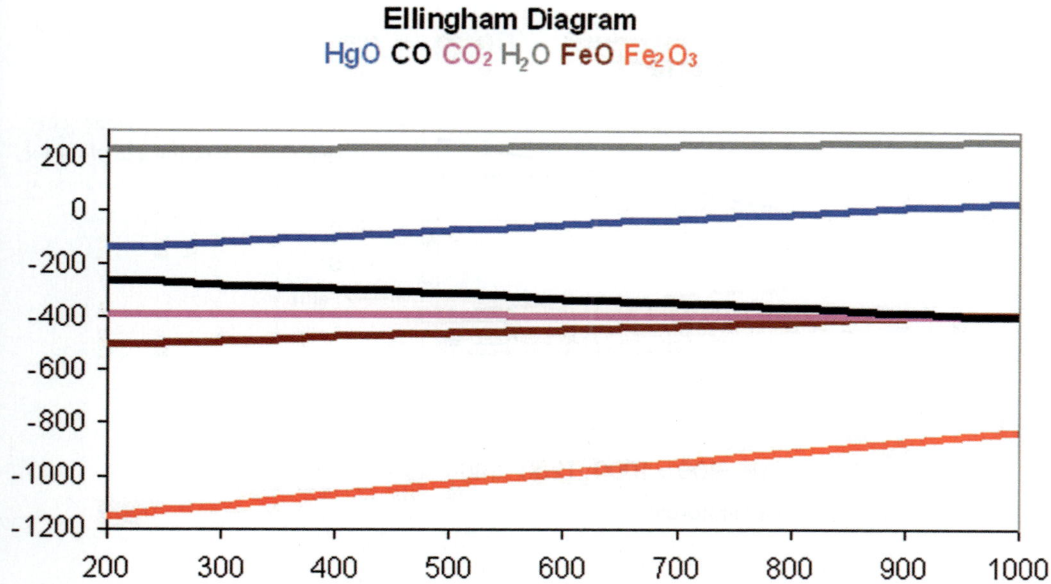

Figure 4. Ellingham diagram for the chemical reaction of mercury, carbon, hydrogen and iron with oxygen. The molar Gibbs energy of reaction [/kJ mol^{-1}] is plotted against reaction temperature (/K)

First and Second Laws ΔG_m ('available heat content') = $\Delta H_m - T\Delta S_m$

Without changing the formal thermodynamic description of chemical reactions, we can readily redefine the Gibbs state function and give it, and its two components, names consistent with the caloric theory of 'heat' and phlogiston theory of reactions of the 18th century. The change in Gibbs energy for an oxidation is referred to as the chemical potential of oxygen within the material relative to its oxide.

We will simply alter the names and symbols of the various energies and revert to specific rather than molar quantities (since phlogiston preceded the chemical-amount balance discovered in 1805 by Dalton). For reactions of an element with oxygen by chemical equation (1), there will be changes in the state functions for which we introduce the names 'entropic', 'caloric' and 'phlogiston' (and reverse the signs so that phlogiston content is a positive concept)

Entropic: $\Gamma = T\Delta S_m/(xM)$
Caloric: $C = -\Delta H_m/(xM)$
Phlogiston: $\Pi = C + \Gamma = -\Delta G_m/(xM)$ [as given above]

then, in phlogistonic/caloric terms

$$\text{X} = \text{calx of X} + C\ (\text{caloric}) + \Pi\ (\text{phlogiston})$$

i.e., caloric and phlogiston are produced when X goes to its calx. $-\Delta G_m/(xM)$ ($=\Pi$) is now regarded not as the propensity of equal masses of X to react with oxygen but the propensity of equal masses of X to produce phlogiston, or as *the phlogiston content of X*.

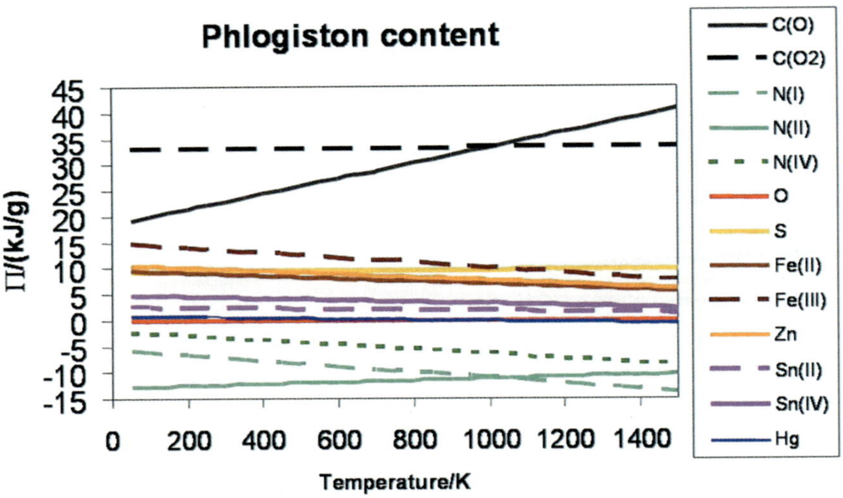

Figure 5. The phlogiston content of some elements calculated as a thermodynamic state function and plotted against temperature at constant pressure

We can now proceed to compute the phlogiston content of any element or compound or mixed material with respect to any of its oxidised states. Having done so, we can then plot the phlogiston content as a function of temperature for all elements in the spirit of Ellingham.

The values of the phlogiston content seen in Figure 5 are quite interesting, and reveal that the general theory of phlogiston was in many ways a reasonable description of the driving force of reactions with oxygen before the advent of chemical thermodynamics and the Gibbs nomenclature of chemical potential.

The first observation that we make is that pure oxygen has no propensity to react with itself and its phlogiston content is zero at all temperatures. Oxygen is 'dephlogisticated air'!

Secondly we note that the phlogiston content of the element hydrogen is so great (above 200) that it is off the diagram. Some of the 18th century phlogistonists, including Priestley at one point, actually believed that pure hydrogen could be phlogiston; such a misconception is now quite understandable.

We also note that the phlogiston content of mercury, unlike most other combustibles, is also very close to zero; in fact it is positive up to 900 K and then it crosses zero and goes negative. It is this simple fact that enabled Joseph Priestley to discover pure oxygen in 1774. By heating the oxide of mercury above 900 K, he was able to eliminate the "phlogiston content" and produce "dephlogisticated air" plus pure mercury.

Conclusions

In a Priestley publication of 1796, which he pointedly addressed to the list of French antiphlogistonists, Priestley wrote as follows in his conclusion.

... The phlogiston theory is not without its difficulties. The chief of them is that we are not able to ascertain the weight of phlogiston, or indeed that of the oxygenous principle. But neither do any of us pretend to have weighed light, or the element of heat, though we do not doubt but that they are properly substances, capable by their addition, or abstraction, of making great changes in the properties of bodies, and of being transmitted from one substance to another.

In conclusion, we can say that although the antiphlogistonists might have rightful claim to be the founding fathers of *inorganic chemistry*, they did not address the question of why chemical reactions take place. They ridiculed the phlogiston theory but had no replacement. We can now see that the reason Priestley adhered stubbornly to the theory until his dying day was because he instinctively knew that there had to be something to account for the propensity of elements to react with oxygen, and the variations in that propensity from one to another. Priestley, it might be argued, was not the last phlogistonist but the first physical chemist. One might speculate that if J. W. Gibbs had been a chemist, rather than a mathematician and engineer, he might have given some credence to the phlogiston theory.

Gibbs theory, as it was originally written (see collected works J. W. Gibbs 1906), was not easy for the uninitiated to understand his great work, and it was 10 years before it was recognised as such. That might be another reason why historians of science have not been so kind to Priestley and his phlogiston. They themselves may not understand Gibbsian thermodynamics, or why chemicals react!

Literature cited

Joseph Priestley "Experiments and observations on different kinds of air", J. Johnson, London (1774)

Antoine Lavoisier "Memoir on combustion in general" French Philosophical Society Journal (1775)

J. Willard Gibbs, "The Collected Scientific Papers Volume I Thermodynamics" Longmans Green (London and New York, 1906)

H. J. T. Ellingham, Journal of the Society of Chemical Industry, pp. 125–133 (May 1944)

Joseph Priestley, "Considerations on the Doctrine of Phlogiston and the Decomposition of Water" Printed by Thomas Dobson (Philadelphia, 1796)

SECTION 11

BOC Symposium: Biological Sciences

BIOLOGY BEFORE DARWIN

Sandra Knapp
Natural History Museum: London

Sandra Knapp

What is biology? Today most people think of it as molecular biology or gene sequencing – but this was not always so. If we can step back a minute and think of biology in its broadest sense we can see that it essentially is the effort to know about the world around us, and is an integral part of the human condition. In the Judaeo-Christian creation myth the first thing Adam did was to name the animals – making biology, or at least a branch of it, truly the oldest profession! As the world grew for human beings, they encountered more and more that they did not understand or even know.

The earliest explorers of the New World speculated that there must have been two Gardens of Eden; the things they were seeing were so very different from those they were used to. The medieval herbalists were among the first to begin to document the diversity of the world around them – after all, it was essential to cure most human ills. Exploration became the way in which European society learned about the world 'out there', but sometimes the reality of what was out there was just too much to believe – for example, the mad, bad William Dampier, an English pirate who in 1679 accidentally circumnavigated the globe and 'discovered' Australia, made natural history observations about the flora and fauna of the southern lands that were, quite frankly, thought to be the ravings of a lunatic.

The 17th century brought an explosion of knowledge about the world outside Europe's borders – this, in part, was responsible for the dawning of the age of enlightenment. In this short commentary to accompany the celebrations of the bicentenary of Joseph Priestley, I will highlight a few personalities – these are not the only people who influenced biology during Priestley's lifetime, but for me represent and illustrate the biology of Priestley's day and the impact of revolution and political upheaval on careers and the perception of science at the time. In a way, we are having a peek into the rich cabinet of the times – and will see how the biology of Priestley's time made the Darwinian revolution possible.

Hans Sloane (1660–1753) – for whom Sloane Square in London is named – was a fashionable physician and founding member of the Royal Society. In the late 17th century he went to Jamaica as the physician to the Governor General, and while there he indulged in his passion for natural history. But, unlike Dampier, he brought back not just stories of far-flung lands, but specimens, actual evidence of what grew there and how it was used. In 1705, he published his first volume of his natural history of Jamaica with the amazing (at least to us in the 21st century!) title of *A voyage to the islands Madera, Barbados, Nieves, St. Christophers, and JAMAICA, with the natural history of the trees, herbs, four-footed beasts, fishes, birds, reptiles etc. of the last of those islands*, ... etc., etc., etc.

Sloane's book documented diversity and its uses – including the use of chocolate, which under the brand name 'Sir Hans Sloane's Milk Chocolate' became a cure-all for all human ills, from tuberculosis to fainting fits. In 1753, Sir Hans Sloane died, leaving his collections to the Nation – with the proviso that a national museum be built in which to house them. This was the birth of the British Museum, of which the Natural History Museum, my own institution, was at one time a part.

Just after the publication of the first volume of the *Natural History of Jamaica* was born the first of my personalities – Carolus Linnaeus or Carl von Linné (1707–1778). In rural Sweden as a young man destined for the clergy, his insistence on collecting and observing plans drove his mother mad. In his early twenties, he made an epic journey to Lapland – even more epic in his telling of the tale – where he began to think about naming plants. He drew wonderful sketches of plants he saw – likening some to figures of Greek mythology. But upon his return, he found he needed to make his name in the wide world, so he went to Holland to get a medical degree. Now, one must remember that in Linnaeus's day, medicine was botany and vice versa, so this was logical. Linnaeus received his degree in record time, two days – he was something of a prodigy.

In Holland, Linnaeus had many supporters and patrons, among them Johannes Burmann. Burmann had a rich and influential friend, George Clifford, who really wanted Linnaeus to come to catalogue the plants in his garden, among the finest in Europe and packed with exotics from all over the world. Eventually Burmann was convinced to part with Linnaeus, ironically with the exchange of a copy of Sloane's *Natural History of Jamaica*! Linnaeus catalogued the plants of Clifford's garden (see www.nhm.ac.uk/botany/databases/clifford) where he described, for the first time, a banana growing in Europe. While working with Clifford, Linnaeus met George Ehret, a rising botanical artist, with whom he devised the diagram of the sexual system – where by counting the numbers of male and female parts, anyone could begin to classify plants.

This made biology – at least botany – accessible to anyone, in fact, it was thought dangerous that women might look at the sexual parts of flowers –

today one wonders why the fuss? By 1753 – coincidentally the year of Sloane's death – Linnaeus was back in Sweden and publishing his seminal work *Species Plantarum* – a complete catalogue of the world's plants known at the time, some 9000 species. This book changed the way people named plants forever – in it Linnaeus introduced his system of trivial names – a two-word abbreviation for a scientific name that has today become the binomial system of scientific naming. Before this time plants and animals were given long polynomial names – essentially phrases in Latin. But as more and more types of plants and animals became know to Europeans, the phrases became longer and longer in order to differentiate among the types of things known.

Linnaeus had the brainwave to use what he called trivial name – a one word designation in addition to the genus name, to aid in memorization of names – after all, it was thought at the time that a good practitioner would hold in his memory the names of all things. So he 'invented' what today we call the binomial system of naming; each thing (or species) has a name of two words – a genus and a species (which started out as Linnaeus's trivial name) – for example, *Tyrannosaurus rex* or *Solanum tuberosum*. Linnaeus was incredibly successful in his own day – he had many students, who went all over the world (some dying on their journeys) to look for plants and to further Linnaeus's own aims. Per Kalm, for example, travelled to North America – Linnaeus named the beautiful North American genus *Kalmia* in his honour.

Joseph Banks, the longest ever serving President of the Royal Society and a prominent 'amateur' botanist, was a keen supporter of the Linnaean system. When he accompanied the voyage to observe the Transit of Venus led by Captain Cook, he took along a student of Linnaeus's, Daniel Solander. Together with their artist, Sydney Parkinson, they saw amazing things – they showed that Dampier's wild imaginings were actually the truth – kangaroos did exist and the flora of Australia was more amazing than even Dampier described. Banks described the kangaroo as " ... an animal as large as a greyhound, of a mouse colour and very swift" – pretty close, and quite poetic! New knowledge, however, was not confined to that coming from the Southern Hemisphere – North America was an unknown land and the newly resident Americans were keen observers of the world around them. Among these were the Bartrams, father John and son William, whose natural history drawings and works were a source of pride to both the English and the new Americans. *Franklinia alatamaha* – named for the great Republican Benjamin Franklin, friend of both the Bartrams – was found by them along the Altamaha River in Georgia, and was last seen in the wild in 1803. Today, all the plants we cultivate are the progeny of William Bartram's first introduction.

The next personality I would like to highlight is the Darwin in biology before Darwin – a man who actually met and was friends with both Benjamin Franklin and Joseph Priestley. Erasmus Darwin (1731–1802) was, like Sloane and Linnaeus, a physician and a country doctor – he was also a keen inventor of an amazing variety of mechanical devices, such as the 'electrostatic coupler' used to alleviate muscle pain. He predicted accurately the future importance of electricity, much against prevailing wisdom of the time. He loved nature and wrote a great deal about it; it is easy to read too much into what Erasmus Darwin wrote, seen in the light of his grandson Charles's accomplishments.

In *Phytologia or the philosophy of agriculture and gardening* published in 1800, for example, he wrote, "The stronger locomotive animals devour the weaker ones without mercy. Such is the condition of organic nature! whose first law might be expressed in the words, 'eat or be eaten', and which would seem to be one great slaughterhouse, one universal scene of rapacity and injustice." (p. 556) – a remarkably modern idea. With Matthew Boulton he founded the Lunar Society – the home of the Lunar Men – whose members supported antislavery ideas, the American and French Revolutions and were so influential in Joseph Priestley's life before he left Britain for America. Darwin wrote many books, and these sparked incredible controversy – William Paley's *Natural Theology* (written in 1802, where the rather famous God the watchmaker was invoked) was thought to have been written in response to Darwin's *Zoonomia* (1794–1796). People objected to his observations on micro-organisms – today we think ridiculous the idea of spontaneous generation of organic from inorganic matter – but in Darwin's day his emphasis on the importance of these "insignificant creatures" was equally reviled.

Darwin's epic poem the *Temple of Nature* was published posthumously, and although it was "intended to amuse"; it aroused considerable passion and vehemence, not least from Priestley himself. Priestley's review of the book stressed his abhorrence of the atheist view taken by Darwin; he could not countenance a world where all beings were interconnected or one in which monkeys, through gradual change in use of digits, became men – Darwin wrote " … mankind arose from one family of monkeys on the banks of the Mediterranean; who accidentally had learned to use the adductor pollicus … and that this muscle gradually increased in size, strength and activity, in successive generations; and by this improved use of the sense of touch, that monkeys acquired clear ideas, and gradually became men." This observation was in a footnote – the rest of the book is written as a poem, and quite amusing it is too!

> Hence on green leave the sexual pleasures dwell,
> And Loves and Beauties crowd the Blossom's bell;
> The wakeful Anther in his silken bed
> O'er the pleased Stigma bows his waxen head;
> With meeting lips and mingling smiles they sup
> Ambrosial dewdrops from the nectar'd cup;
> Or buoyed in air the plumy Lover springs,
> And seeks his Bride on panting Hymen wings.

Flower sex still had the power to titillate. The footnote to this stanza, interestingly, refers to oxygenation of vegetable blood, another interesting Priestley connection. But the ideas in the verse were radical, too radical for many.

Our last personality is usually mentioned in the same breath as Charles Darwin – Erasmus's grandson, but in fact, he was a contemporary of Erasmus Darwin and Joseph Priestley, **not** Charles Darwin. He is usually treated as the straw man to show Charles Darwin's true genius, a completely unfair juxtaposition. Jean Baptiste Pierre Antoine de Monet, Chevalier de Lamarck (1744–1829) – better known as Jean Baptiste de Lamarck, or just Lamarck – was a member of a minor French aristocratic family, and as befitted a younger son, went into a career in the military. After distinguishing himself in battle, he suffered an accident, and left the army.

During a trip to Monaco, Lamarck became fascinated by natural history, particularly botany. He began to study the subject (while working as a bank clerk) and came to the attention of the pre-eminent French naturalist of the day, Georges-Louis Leclerc, Comte de Buffon. Buffon, interestingly, was vehemently anti-Linnaean, setting up a European rift in science that persisted for decades – but that story is not for telling here. Lamarck wrote a manual of the plants of France, the first ever attempted, and Buffon arranged for its publication and extolled its virtues.

Lamarck's 1778 *Flore Française* – funnily enough published the year Linnaeus died – was a scientific book with a difference. In it he introduced the use of common names for plants (rather than the Latin scientific names in vogue at the time, again a step in making science available to everyone) and he produced for the first time a key for identification that involved two-way choices, a dichotomous key – today commonplace, but then a novelty. Buffon employed Lamarck as tutor to his son, thus ensuring his place in the scientific establishment at the Jardin du Roi in Paris. But Buffon died on the eve of the French Revolution, and the community of French scientists split, not along revolutionary lines, but along "Buffonite" and "anti-Buffonite" (*i.e.*, Linnaean) lines. Amazing.

Lamarck, a disciple of Buffon and of an aristocratic family, survived the upheaval of the French Revolution and in the 1793 reorganization of the Jardin du Roi into the Museum d'Histoire Naturelle, he was appointed curator of worms – a definitely second class job. Worms were not thought to be interesting; prestige was attached to the study of vertebrates or "higher" animals. But Lamarck rose to the challenge, and totally revolutionized our understanding of these fascinating animals. His knowledge of them prior to his appointment was confined to shells, but through careful and diligent study he reclassified the 'invertebrates' – a term he invented – into seven classes; all of which are still in use today. Before Lamarck, naturalists did not recognise the crustaceans (crabs and allies) as distinct from arachnids (spiders), there were just worms and not worms. He emphasized the interconnections of all organisms and identified what today we call homologies – structures resulting from descent from common ancestors. He saw life as organised in a series of increasing complexity, progressing from the simple to the complex, in contrast to the prevailing view that simpler creatures were the result of degradation and thus inferior. His tree diagrams drawn to show this scale of complexity form the basis for the paradigm in use today, albeit with radically different philosophical roots. Like with Erasmus Darwin, it is easy to see in Lamarck's ideas conceptions far ahead of their time, but it is important to remember that we see these ideas through evolutionary glasses – we have the advantage of knowing that Charles Darwin articulated the mechanism the earlier generations were searching and our generation can see glimmers of it in much of the work of his predecessors. For all the good ideas he had, Lamarck had some really off-the-wall ones too. He thought, for example, that life regenerated itself constantly from primordial ooze; he set up an alternative chemistry to that of Lavoisier; he argued with almost everyone. In fact, he was probably his own worst enemy.

Lamarck is remembered, however, not for his taxonomy of invertebrates, nor his invention of the term *biology*, nor his monumental *Flore Française*, but instead for his two "laws" from *Philosophie zoologique* (1809) which became enshrined as the anti-Darwinian doctrine of "inheritance of acquired characteristics". These laws were as follows:

First Law: 'In every animal which has not passed the limit of its development, a more frequent and continuous use of any organ gradually strengthens, develops, and enlarges that organ, and gives it a power proportional to the length of time if has been so used; while the permanent disuse of any organ imperceptibly weakens and deteriorates it, and progressively diminishes its functional capacity, until it finally disappears.'

Second Law: 'All the acquisitions and losses wrought by nature on individuals, through the **influence of environment in which their race has long been placed**, and hence through the influence of the predominant use or permanent disuse of any organ; all these are **preserved by reproduction to the new individuals which arise**, provided that the acquired modifications are common to both sexes, or at least to the individuals which produce the young.' (bold face mine)

Lamarck articulated these principles in order to explain the well-known phenomenon of what was perceived as the fitness for purposes of structures in nature – birds that lived on shorelines had long legs that allowed them to wade to hunt for food; plants growing in dry places were smaller and flowered more rapidly than those in wetter, more benign places; giraffes had long necks that allowed them to browse where others could not reach; etc. His laws are caricatured as, to paraphrase – "giraffes stretch their necks, their children get long necks too". Reading what Lamarck actually said though, shows us this is not what he meant. He emphasized the role of generations (read to 21st century eyes *heredity*) and environment (to 21st century eyes *adaptation*) – not so strange after all. But even though he got close, Lamarck didn't really articulate the real mechanism for organic change, but he was getting there, as was Erasmus Darwin. These biologists of Priestley's day were looking for a mechanistic explanation for the changes they saw in the world around them, they had rejected the creationist view and were searching for ways to explain what they observed – just as Priestley searched for ways to explain his various types of air. In 1797 Lamarck stated, rather dangerously for his time, "life ... is a physical fact, albeit somewhat complex in its principles". By the time he was writing *Philosophie zoologique* in 1809 he was even more definite – "It seems, as I have already said, that time and favourable circumstances are the two principal means which nature uses in creating all its products. We know that time has no end for it, and is consequently always at its disposal." – in other words, no superior being or God is necessary, time and favourable circumstances are all that are needed. Dangerous sentiments to express publicly at the time when Napoleon and the Vatican had signed a concordat increasing the power of the Church in French life.

Much of how Lamarck is remembered in the English-speaking world stems from the *Eloge* written after his death by the great French anatomist Georges Cuvier. These written eulogies were used to either sanctify or assassinate colleagues – Cuvier used his to cut Lamarck, who died blind and essentially penniless, down to size as the mediocrity Cuvier thought him to be – the piece is subtle and cutting, and set Lamarck up as a person whose "indulgence of a lively imagination has led to results of a more questionable kind". Pretty nasty stuff, especially from one as intelligent and respected as Cuvier. Lamarck deserves to be remembered as part of the panoply of 18th century science, not as a textbook caricature set up in opposition to the science of the next century.

Exploration of the world carried on apace, and scientists continued to seek explanations for why the world was the way it was, and how it got that way. One of the many, many men who went on a voyage of discovery was a young Charles Darwin, grandson of Erasmus. A paying passenger-naturalist on the HMS Beagle, Darwin the younger visited South America, where he collected specimens and observed with fascinated eyes a world quite different from that of the English shires. On his return, he retired to the country and began to think – and ultimately, with a little push from another explorer, Alfred Russel Wallace, he articulated the mechanism his predecessors were looking for – evolution by natural selection. In doing so, he changed the way in which we see our world, and our place in it, forever.

His visionary leap was made possible by the development of biology during Priestley's generation – the themes of revolution, change and the adherence to materialist principles made it possible for Charles Darwin to come to natural selection as the driving force for evolution. Their ideas and searchings were as revolutionary and evolutionary as biology post-Darwin – they may seem crazy today, but think of what our descendents will say about us!!

Further reading

(This list is by no means complete, the amount published by and about these eminent men and their times is amazing, and all of it opens a fascinating window on scientific life at Priestley's time.)

Blunt, W. 2001. *The Compleat Naturalist: a Life of Linnaeus*. 2nd edition. Frances Lincoln, London.
Brooks, E.St. John. 1954. *Sir Hans Sloane: the Great Collector and his Circle*. Batchworth Press, London.
Burkhardt, R. W. Jr. 1995. *The Spirit of System: Lamarck and Evolutionary Biology*. 2nd edition. Harvard University Press, Cambridge, Mass., London.
Corsi, P. 1988. *The Age of Lamarck: Evolutionary Theories in France 1790–1830*. Revised and updated, translated from the Italian by J. Mandelbaum. University of California Press, Berkeley, Los Angeles, London. (First published in 1983 as *Oetre il mito: Lamarck e le scienze naturali del suo tempo*. Il Mulino, Bologna).
Dampier, W. 1998. *A New Voyage Around the World: the Journal of an English Buccaneer*. Hummingbird Press, London. (First published in 1697 as *A new voyage around the world*. James Knapton, London).
Darwin, E. 1794–1796. *Zoonomia; or, the Laws of Organic Life*. In three parts (I, 1794, II & III, 1796). J. Johnson, London.
Darwin, E. 1803. *The Temple of Nature or, the Origin of Society: a Poem, with Philosophical Notes*. J. Johnson, London.
Darwin, C. 1859. *On the Origin of Species by Means of Natural Selection or the Preservation of Favoured Races in the Struggle for Life*. John Murray, London.
De Beer, G. R. 1953. *Sir Hans Sloane and the British Museum*. Oxford University Press, Oxford.
King-Hele, D. 1999. *Erasmus Darwin: a Life of Unequalled Achievement*. Giles de la Mare Publishers Ltd., London.
King-Hele, D. (ed.) 2003. *Charles Darwin's* The life of Erasmus Darwin. Cambridge University Press, Cambridge.
Koerner, L. 1999. *Linnaeus: Nature and Nation*. Harvard University Press, Cambridge, Mass.
Lamarck, J. B. 1778. *Flore Française or Descriptions Duccintes de Toutes les Plantes qui Croissent Naturellment en France, Disposes Selon une Nouvelle Méthode d'Analyse*. 3 vols. Imprimerie Royale, Paris.
Lamarck, J. B. 1984. *Zoological Philosophy; an Exposition with Regard to the Natural History of Animals*. University of Chicago Press, Chicago (translated from the French by Hugh Elliot; originally published in 1809 as *Philosophie zoologique*).
Magee, J. in press. *The Art and Nature of William Bartram*. Penn State Press, Phildelphia.
Rice, T. 2000. *Voyages of discovery: three centuries of natural history exploration*. Scriptum Editions & the Natural History Museum, London. (see also http://www.nhm.ac.uk/museum/tempexhib/voyages/index.html).
Stafleu, F. A. 1971. *Linnaeus and the Linnaeans: the Spreading of their Ideas in Systematic Botany, 1735–1789*. A. Oosthoek's Uitgevermaatschappij N. V., Utrecht.

DR. PRIESTLEY AND THEORIES OF THE BRAIN

Elfed Huw Price
Wellcome Institute for the History of Medicine
University of Oxford

Elfed Huw Price

Introduction

Joseph Priestley (1733–1804) is most commonly remembered today for his part in the 'discovery' and isolation of oxygen (or 'dephlogisticated air'), and as a radical religious and political reformer who had close ties with the founding fathers of the United States of America. Yet he is generally overlooked for the significant role that he played in pioneering a physiological concept that is generally taken for granted today: namely, the notion that the mental processes are dependent upon the structure and function of the brain.

Throughout Priestley's lifetime it was heretical to suggest that the brain might 'think'. Anglican dogma strictly denied that mind could be produced by, or dependent upon, matter. *Thought* was a property of the *soul*, an immaterial substance that was the very antithesis of matter. Thus, whilst philosophers favoured the brain as the most likely seat of the soul, the cerebral organ itself was not deemed capable of thought. The brain was thought to act as a distilling device, the function of which was to refine 'animal spirits' from the blood and propagate them around the body via the nerves.[1] These spirits, and the nerves that contained them, mediated between the contradistinct substances of soul and body.

In the latter 1770s, Priestley tackled dualism head on, publishing several works that argued that there was no soul distinct from matter, and that it was the organisation of the brain which was responsible for sentience. In doing so, he incurred both the wrath and ridicule of more orthodox contemporaries. The aim of this paper is to explore the religious context of Priestley's adoption of 'materialist' theories of the brain in the last three decades of the eighteenth century.[2]

Religion and science were inseparable for Priestley, as they were for many of his contemporaries. Historians of science and religion have imposed a divide between these two aspects of his life: one of the aims of this paper is to re-integrate the two in Priestley's individual thought, and through doing so, hopefully potentially give some insight into popular beliefs of the era.

Priestley's Religious Background

Priestley was brought up a Calvinist and during his youth accepted the conventional division between the substances of immaterial body and material soul. However, subsequently, his researches into science and theology led him towards increasingly materialistic doctrines of the soul and mind. The transition from orthodox dualism to monist materialism took place through two 'conversions' in his religious beliefs, in a process that spanned the greater part of his life.

The first of these transformations in religious affiliation took place at the non-conformist Daventry Academy, which he attended in his late teens and early twenties. During this time, the academy was a hotbed of religious debate. As he Priestley later recalled: "In my time, the Academy was in a state peculiarly favourable to the serious pursuit of truth, as the students were about equally divided upon every question of much importance, such as liberty and necessity, the sleep of the soul, and all the articles of theological orthodoxy and heresy; in consequence of which, all these topics were the subject of constant discussion ... I saw reason to embrace what is generally called the heterodox side of every question".[3]

It was at Daventry that Priestley became acquainted with David Hartley's systematised 'physiology' of mind, in *Observations on Man, His Frame, His Duty and His Expectations* (1749). Hartley attempted to relate the processes of the mind to those of the body, and particularly of the brain and nerves.

The Observations offered a sensory-based explanation for the development of ideas in the mind. Towards the end of the previous century, John Locke had proposed that all complex ideas were built up from simpler ideas, and that these were ultimately derived from the senses. Yet Locke had specifically avoided speculating as to the physiological basis of the process.[4] Hartley adopted Locke's system, and 'anatomised' it in the bodily substrate in the nerves and brain.

According to Hartley's theory, sensations excited vibrations of minuscule particles in the nerves at the point of contact. These were propagated to the medullary substance of the brain (*i.e.*, the inner part, as distinct from the cortex), the region of the brain he supposed to be materially involved with the processes of the mind.[5] Different qualia (*i.e.*, type of sensory experience) were induced depending on the quality of the vibration. Here Hartley drew upon musical terminology to describe these, postulating that vibrations could vary in terms of tone, pitch and volume, in addition to which region of the brain was affected. When two or more simple vibrations were presented to the brain at the same time a material alteration of the medullary substance would bind these together by 'association'. Over time, this then would form the material basis for the development of complex ideas from simple ones.

Despite his intricate emphasis on the inter-relationship between the processes of mind and body, Hartley was no monist. The opening lines of his *Observations* make this clear:

> Man consists of two parts, body and mind. The first is subjected to our senses and inquiries, in the same manner as the other parts of the external material world. The last is that substance, agent, principle, &c. to which we refer the sensations, ideas, pleasures, pains, and voluntary motions.[6]

The impact of this work on Priestley's thought should not be underestimated. Priestley accredited it with producing "the greatest, and in my opinion, the most favourable effect on my general turn of thinking through life".[7] It was upon reading Hartley that Priestley came to believe that human behaviour was subject to the laws of necessity. It seems highly likely that Hartley also contributed to Priestley's shift towards Arianism during this time.[8] In later years, Hartley's work would also serve as the premise for much of his writing on the sentience of the brain.

For the next twenty years, Priestley remained an Arian, retaining belief in an immaterial sentient principle, but one that could not function without a body.[9] However, during the 1770s, he underwent a second transition of faith, in which he shifted from the tolerated heresy of moderate dualism to the extreme heresy of monist materialism. It is to this transition that we now turn.

Priestley and Theories of the Brain

During an intellectually significant three-year period (between 1775 and 1778) Priestley abandoned interdependent dualism in favour of the notion of the thinking brain. By 1775 Priestley was already well established in the public domain. Numerous editions of his works were in print on diverse scientific topics such as electricity, optics and chemistry, in addition to explicitly theological writings. From amongst his works first published in this year it is possible to find two distinct (yet mutually compatible) hypotheses concerning the functions of the brain. Both were indebted to previous traditions; yet Priestley presented his own distinct interpretation of each.

One of these theories was raised in the "Queries, Speculations and Hints" section of Priestley's *Experiments and Observations on Different Kinds of Air.*[10] Here he posited a physiological model of respiration, based on the cycle of energy bearing phlogiston.

> Let it be considered that the proper nourishment of an animal body, from which the source and materials of all muscular motion must be derived, is probably some modification of phlogiston. Nothing will nourish that does not contain phlogiston, and probably in such a state as to be easily separated from it by the animal functions ... it should seem that the phlogiston which we take in with our aliment, having discharged its proper function in the animal system (by which it probably undergoes some unknown alteration) is discharged ... by the lungs.[11]

Drawing on his knowledge of electricity and its effects on the body, Priestley speculated that it was not 'raw' phlogiston that caused muscular motion, but a particular 'electrical' modification of it, carried out in and by the brain.

> My conjecture ... is, that animals have a power of converting phlogiston from the state in which they receive it in their nutriment, into that state in which it is called the electrical fluid; that the brain, besides its other proper uses, is the great laboratory and repository for this purpose; that by means of the nerves this great principle,

thus exalted, is directed into the muscles, and forces them to act, in the same manner as they are forced into action when the electric fluid is thrown into them *ab extra*.[12]

This portrayal of the animal economy was not completely at odds with the established one: it retained the concept of the brain as secreting gland, the fountainhead of the nervous system. The theory deviated from that of Willis is in the substitution of electricity in the place of animal spirits, and by combining it with a phlogiston theory of respiration.

Although this theory of the brain attempted to explain muscular motion in electrochemical terms, the passage did not openly challenge dualism; moreover, the focus was on motion, not the mind. Whilst he hinted that the brain had "other proper uses", Priestley left open what these might be. He dealt with this topic in a separate series of publications dating from 1775 to 1778 in which he presented a monist theory of mind and matter, based on the work of David Hartley. This model was both more polemical and more controversial than the first; it was in this theory that he first suggested, and subsequently proclaimed, that it is the brain that thinks.

The first of these texts, published in 1775, was an edition of Hartley's *Observations*. In his memoirs, Priestley recalls that his re-publication of this work was prompted by a disagreement with the (immaterialist dualist) Scottish school of common sense philosophy.[13] In the work itself he states that the intent of his publication was to simplify the work to allow it a more general reception.[14] Hartley's *Observations* had made some impact in its first edition, but had not received the recognition Priestley felt it deserved.

The 1775 Edition of the Observations

The main body of Priestley's 1775 edition was essentially the same as the first volume of the original. Priestley had made minor changes to this work, notably his attempt to relocate references to the controversial theory of vibrations to the footnotes, to avoid difficulties readers might have with accepting Hartley's theory purely on this account.[15] Priestley's most controversial contribution was not any alteration within Hartley's work itself, but a seemingly innocuous comment contained within the *Introductory Dissertations* he attached. Whilst Hartley had taken care to stress the duality of immaterial mind and material body, Priestley openly speculated that this immaterial principle might not be unnecessary and superfluous to the system. Echoing John Locke's famous comments on 'thinking matter', Priestley initially couched his suggestion cautiously as mere speculation:

> We may, by this time, I should think, be prepared to admit the *possibility* of a mass of matter like the brain, having been formed by the almighty creator, with such exquisite powers, with respect to vibrations, as should be sufficient for all the purposes above-mentioned ... And it is only the bare *possibility* of the thing that I now contend for.[16]

Yet, for the majority of his contemporaries, even the *possibility* of materialism was objectionable enough. As Priestley later recalled in his Memoirs:

> In one of these [introductory dissertations] I expressed some doubt of the immateriality of the sentient principle in man; and the outcry that was made on what I casually expressed on that subject can hardly be imagined. In all the newspapers, and most of the periodical publications, I was represented as an unbeliever in revelation, and no better than an Atheist.[17]

Far from than quashing Priestley's voracious appetite for the 'truth', this response only served to spur him on. The public outcry: " ... led me to give the closest attention to the subject, and the consequence was the firmest persuasion that man is wholly material ... I therefore digested my thoughts on the subject, and published my *Disquisitions relating to Matter and Spirit*, also the subjects of 'Socinianism' and 'Necessity', being nearly connected with the doctrine of the materiality of man".[18]

In addition, Priestley compiled his correspondence that resulted from the controversy, the bulk of it being that between Priestley and his dissenting friend Richard Price. This was published in 1778 as yet another volume: *A Free Discussion of the doctrines of Materialism and Philosophical Necessity*. By the *Free Discussion* of 1778, his statements concerning the role of the brain in the processes of thought are far from his cautious speculation of 1775:

> I infer that the business of thinking is wholly carried on in, and by the brain itself, because all the effects from which we infer the faculty of thinking can be traced to the brain, and no farther.[19] And ... In my opinion, there is just the same reason to conclude that that brain thinks, as that it is white, and soft.[20]

Within three years of his edition of Hartley's opus he had publishing a small corpus of work defending the interconnected doctrines of materialism, necessity and Socinianism.[21]

Of all his writings, it is surely within these that we might find the clearest explanations of how his religion, philosophy and science were interwoven.

The Disquisitions

In *Disquisitions Relating to Matter and Spirit* (1777) Priestley drew together religion, history and science to argue that the very distinction between matter and spirit is invalid. The frontispiece indicated the general intentions, with a citation from the Swiss naturalist, Charles Bonnet, in support of the possibility of materialism.[22] Priestley acknowledged that his conceptualisation of matter was indebted to similar theories that were developed independently by the Croatian Jesuit, Roger Boscovich (1711–1787) and the English astronomer John Michell (1724–1793).[23] Priestley had adopted the Boscovich-Michell theory of matter several years earlier. In his *Histories of Discoveries relating to Vision, Light and Colours* (1772) he related how:

This scheme of the *mutual penetration of matter*, first occurred to Mr. Michell on reading *Baxter on the Immortality of the Soul*. He found that this author's idea of matter was that it consisted, as it were, of bricks cemented together by an immaterial mortar. These bricks, if he would be consistent in his own reasoning, were again composed of less bricks, cemented likewise by an immaterial mortar, and so on *ad infinitum*. This putting Mr. Michell upon the consideration of the appearances of nature, he began to perceive that the bricks were so covered with this immaterial mortar, that, if they had any existence at all, it could not possibly be *perceived*, every effect being produced by at least in nine instances in ten certainly, and possibly in the tenth also, by this immaterial, spiritual and penetrable mortar.

Instead, therefore, of placing the world upon the giant, the giant upon the tortoise, and the tortoise upon he would not tell what, he placed the world at once upon itself.[24]

The controversy between soul and matter was a debate concerning *substance*, in the sense of 'that which stands under', *i.e.*, the underlying supporting essence, of a thing. According to substance dualism, the essential properties of matter were that it was inert, impenetrable, and extended in space, whilst soul was the converse – essentially active, thinking and without location.

In Priestley's version of materialism, matter was deconstructed into attractive and repulsive forces, and as such did not accord to the conventional understanding. For Priestley, spirit and matter were conflated, and hence (he argued) their properties were rendered commensurable. If matter and spirit were one and the same, there was no reason why this monist substance might not be able to think.

In this paper has less concern with the specific merits or problems of the various dualist or monist positions than with the religious concerns that accompanied them. In the dedication to the *Disquisitions* (addressed to Reverend William Graham, an outspoken free thinking cleric) it is clear that they were primarily intended as a statement of religious belief. More specifically, it was intended as a defence of Socinian doctrine against dualist religious beliefs. Priestley writes,

Whatever you may think of some parts of my reasoning in the principal work, now presented to you, I am confident you will approve of the main *object* of it, and especially the Sequel.[25] You have long been an asserter of the *proper unitarian doctrine*, and cannot be displeased with my endeavouring to trace their source in heathen antiquity, those capital corruptions of Christianity – the *Athanasian* and *Arian* opinions.[26]

As the most prominent materialist in the late eighteenth century Britain, an understanding of Priestley's theories and motivations is invaluable.[27] To get more fully to grips with Priestley's materialism, it will be necessary to clarify how this relates to, and differs from, the Athanasian and Arian systems. We will now therefore turn to the religious debate itself.

Athanasians, Arians, Socinians and the immateriality of the soul.

Priestley, Disquisitions, p. 248

> ... divines (and not philosophers) were first convinced that, though man has a soul distinct from his body, its powers and perception and action depend upon the body, and that the whole state of man is in a state of insensibility from death to the resurrection.

In Priestley's mind, the debates over the materiality of thought were primarily a matter of religious truth. Priestley highlighted the three primary factions with divergent beliefs concerning the soul: Athanasians, Arians and Socinians (also known as Unitarians). Each of these interpretations of Christianity was named after a religious figurehead (Athanasius, Arius and Socinus respectively); in each case, these names refer to advocates and upholders of the respective belief system rather than originators of it. During his lifetime Priestley passed through all three, having being brought up a Calvinist (Athanasian), left Daventry Academy as an Arian, and having fully embraced Socinianism during the 1770s in his early forties.

All too often, historical attempts to delineate between these three sects are reduced to a single point of difference – commonly opinions concerning the divinity of Jesus Christ. But this is an oversimplification. Each of these religious groups held its own philosophical system, its own paradigm, which consisted of a network of beliefs that were internally consistent within each 'sect', but incommensurate between groups.

There are a number of characteristic beliefs that serve to distinguish between the three belief-systems. Several of these relate to the Christian understandings of the divine Trinity. All sects were able to accept the divinity of God as self-evident. To be classified as orthodox Trinitarian is to believe in the equal claims to divinity of the Father, Jesus Christ, and the Holy Ghost. On this basis, traditionally, Athanasians are denoted to be 'Trinitarian' and Socinians to be 'Unitarian', whilst Arians fall somewhere in-between. It was the divine status of Jesus that was usually raised as a point of contention.

Whilst in the eighteenth century the Trinity had an immediate and visceral importance, it is a topic of some obscurity today. Therefore I intend to address the network of beliefs that surround this complex issue via a second doctrine that split the three camps, but one more easily grasped by the modern mind; that of the immortality of the soul. In the eighteenth century one's opinion on the immortality of the soul was "a litmus test for orthodoxy".[28] The doctrine of the immortality of the soul, at least according to Priestley's account, was merely one part of a complex internally consistent system of beliefs, having a direct bearing on understandings of the doctrines of the immaterial soul, the Trinity and of the Atonement. The only one of importance here is its implications for theories of mind and matter.

Athanasianism

The nomenclature of the 'Athanasian' owes its origins to a high profile ecumenical debate in the 4th century AD between two clergymen named Athanasius and Arius.[29] There is no question that the immortality of the soul was a central pillar of the Athanasian system; indeed, the name *Athanasius* itself (from *'a-thanatos'*) is Greek for immortality. This immortality of the soul extended in both temporal directions, from the alpha to omega of time; the soul was in existence both before and after the individual lifetime on Earth.

For the Athanasians, the soul was deemed to be essentially active. For it to be immortal it must continue to exist in an active state both before birth and between death and resurrection. To be able to do so, it must be able to function independent of the material moorings in the body.[30] Consequently, the soul was an active, independent, substance in its own right. (The most common theological focus of this was one's beliefs concerning the 'intermediate state of the soul', or the gap between death and resurrection of the body.[31]) Thus, by believing in a immortal substantial soul, the Athanasians also required belief in an immaterial soul, and one that was in no way dependent on the body for its functions – including that of thought.

Arianism

Arianism was the second system of belief adopted by Priestley. The Arian standpoint can be gauged both from Priestley's own reports, and his conversations with his Arian friend Richard Price. Arians were dualist. Like Athanasians, there was a spiritual component of man that survived the death of the body. Yet, unlike Athanasians, this soul was not fully immortal; it was not self-existent, not an emanation of the divine, but created by *him*. As such it needed a substance to stand upon; it required a body to think. Similarly, between mortal death and resurrection, its state of existence was only partial.[32] Arians believed that the soul slept (and possibly dreamed slightly) between death and resurrection.[33]

Socinianism

The last phase of Priestley's religious odyssey was Socinianism, or Unitarianism, so called because they opposed the doctrine of the Trinity and taught that there was only one god. The name Socinian is of much later origin than the others described above, being initially applied to the followers of the Italian lay preacher, Faustus Socinus (1539–1604), who had emigrated to Poland and built up a religious community there. As such, it is sometimes mistakenly seen this as a younger offshoot of Arianism.[34] Yet, as both Priestley and his critics were well aware, the tenets of 'Socinianism' had much older antecedents, and dating back to the days of the original Athanasian debate and beyond.[35]

The Unitarians denied that Christ was born divine – he was born human, and was 'adopted' as God's son. The absolute humanity of Jesus was mirrored in the absolute materiality of man.

> God made man of the *dust of the ground*, and by simply animating this organized matter, made him that living, percipient, and intelligent being that he is.[36]

For Priestley there was no immaterial soul, and thus no 'thought' separate from the body. Hence, the cessation of functioning of the body also signified the absolute cessation of thought. He makes this position explicit in this passage from the *Free Discussion*:[37]

> I suppose that the powers of thought are not merely suspended, but are *extinct*, or *cease to be*, at death ... I do not ... think that any thing that I have advanced implies that *the soul*, that is, *the man* loses his existence at death, in any other sense than that the man loses his *power of thinking*.

Conclusion

Contemporary use of the term 'materialism' connotes a scientific secular world-view in which only material things are of real influence and worth. In contrast, Priestley's materialism was one in which he merged matter and spirit to establish a scientific and religious Unitarian ideology which he felt held true to reason, science and Christianity.

If we are to understand the actions and thoughts of the past, we must be sympathetic to the differences between past and present. In Priestley's life and work, religious debates on the nature of the soul were fully integrated into the discourse and debate of 'scientific' discovery. The religious controversies were neither idiosyncratic to Priestley, nor restricted to the eighteenth century. The issues he addressed were to emerge again forcefully during the early decades of the nineteenth century, in the phrenological doctrines of Gall and Spurzheim and the writings of London surgeon William Lawrence.

Priestley would have been the first to acknowledge that there are no 'great men'; that he was merely one person in a deterministic system. Nevertheless, modern day psychology owes a greater debt to Priestley than is generally acknowledged. His work made a significant contribution on several counts: in promoting electrochemical understandings of the physiology of the body; in helping to establish Hartley's neural basis for the association of ideas; and in challenging the dogma of the independent immaterial soul. Equally, cultural historians are equally indebted to Priestley for his outspoken honesty and integrity, in that these provide a guide, a Rosetta Stone, that enables them to bridge the divide between histories of religion and science. This may provide a means for gaining greater understanding of doctrines of mind, matter and the brain.

I close with an excerpt from a poem by the young Unitarian, Samuel Taylor Coleridge. This was written in 1794, the same year Priestley left for America to escape the threat of the Church and King lobby, some of whom had destroyed Priestley's home and laboratory two years before. Coleridge, a scholar and philosopher as well as a poet, identifies the lineage of the development of the theory of the "sentient brain", without pigeon holing him as *either* philosopher *or* priest, but acknowledging him as both.

Adoring Newton his serener eye
Raises to heaven: and he of mortal kind
Wisest, he first who marked the ideal tribes
Up the fine fibres through the sentient brain.
Lo! Priestley there, patriot, and saint, and sage,
Him, full of years, from his loved native land
Statesmen blood-stained and priests idolatrous
By dark lies maddening the blind multitude
Drove with vain hate. Calm, pitying he retired,
And mused expectant on these promised years.[38]

NOTES AND REFERENCES

1. The prominent anatomist Thomas Willis was influential in promoting this neurological model in the late seventeenth century, as was George Cheyne in the eighteenth.
2. The term materialism is used here guardedly, as Priestley's theory re-defined matter and spirit as one and the same substance.
3. Priestley's Memoirs, republished in *Autobiography of Joseph Priestley* (Adams & Dart, 1970) pp. 75–76
4. Locke Essay *Concerning Human Understanding* (London, 1689)
5. Hartley's first proposition (based on unspecified findings by similarly unspecified physicians and anatomists) was that: *"the white medullary substance of the brain, spinal marrow, and the nerves proceeding from them, is the immediate instrument of sensation and motion."* It is of note that, for Hartley – and consequently also for Priestley – it was the *white medullary* (or inner) part of the brain that was the seat and source of mental influence. Meanwhile, Hartley implied that the cortex had no role whatsoever in "sensation, voluntary motion, memory and intellect". *Hartley's Theory of the Human Mind on the Principle of the Association of Ideas; with Essays relating to the Subject of it.* (London, 1775): p. 7.
6. Priestley *Hartley's Theory of the Human Mind* (1775): p. i
7. Priestley *Memoirs* (1970) p. 76
8. Priestley became an Arian either during his time at Daventry or immediately after. *ibid.* pp. 77–79.

9. Priestley was still writing from a dualist viewpoint in his *Institutes of Natural and Revealed Religion* (1772): see *Disquisitions Relating to Matter and Spirit* (London, 1777): p. xi.

10. *Experiments and Observations* (1775): pp. 276–278. I am indebted to an article by Simon Schaffer for drawing my attention to this passage. Schaffer, S. "Priestley and the Politics of Spirit."' in Anderson, R G W and Lawrence, C Eds. *Science, Medicine and Dissent: Joseph Priestley* (1733–1804) (Wellcome Trust/Science Museum, London 1987.): p. 47

11. *Experiments and Observations* (1775): p. 276

12. *ibid.*: pp. 277–278

13. Priestley *Memoirs* (1970): p. 113

14. Priestley *Hartley's Theory of the Human Mind* (1775): p. iii

15. There were two significant competing theories of nerve structure and function during the eighteenth century. One held that nerves were hollow, and allowed transmission of some kind of nervous fluid, which was responsible for conveying sensibility and motility from the brain to the organism. This theory echoes the theory of the brain as a gland, secreting animal spirits. The other – as adopted by Hartley – noted that the brain and nerves were made of the same solid substance, and that vibrations of the minuscule particles within the nerves facilitated the propagation of the said vibrations to the brain. The doctrine of vibrations was seen as implying materialism, as it obviated the need for a separate immaterial soul (*e.g.* Priestley, *Hartley's Theory of the Human Mind* (1775): p. xix) Priestley had his own reasons for supporting the alternative fluid theory, as it was part of his electrical/phlogiston theory of the brain (as described above). By ostensibly avoiding this debate, Priestley effectively side-stepped a swathe of potential criticisms of Hartley's theory on specific anatomical grounds. For instance, even before Hartley's original work, Boerhaave had thrown his substantial intellectual weight *against* vibration theory in his 1743 *Institutes*. See Smith, CUM "David Hartley's Newtonian Neuropsychology" J. Hist. Behav. Sci. 1987, 23: 123–136: 131

16. Priestley *Hartley's Theory of the Human Mind* (1775): p. xviii

17. Priestley *Memoirs* (1970): p. 114.

18. Priestley *Memoirs* (1970): p. 114. The second work referred to is *The History of the Philosophical Doctrine concerning the Origin of the soul, and the Nature of Matter*, which was published, appended to the *Disquisitions* (1777). The third is *The Doctrine of Philosophical Necessity* (London, 1777).

19. *Free Discussion* (1778) pp. 49–50

20. ibid.: p. 61. The 'medullary' region is white, not grey. (cf. endnote 6)

21. "The three doctrines of materialism, of that which is commonly called Socinianism, and of philosophical necessity, are equally parts of one system, being equally founded on just observations of nature and fair deductions from Scripture." *Disquisitions*: 356

22. "Si quelqu'un demontreroit jamais, qu *l'ame est materielle*, loin de s'en alarmer, il faudroit admirer la puissance, qui auroit donné a la matiere la capacité de penser. BONNETT [sic]. *Palingenesis*, Vol. I, p. 50."

23. *Disquisitions* (1777): p. xii

24. Cited in Priestley *Disquisitions* (1777): p. 21

25. A second treatise contained in the same volume entitled *The history of the philosophical doctrine of concerning the Origin of the Soul, and the Nature of matter; with its Influence on Christianity, especially with Respect to the Doctrine of the Pre-existence of Christ.*

26. *Disquisitions* (1777) p. vii–viii

27. John Yolton went so far as to claim that Priestley stood alone in his prominent advocacy of materialism in eighteenth century Britain. *Thinking matter: materialism in eighteenth century Britain* (Blackwell, 1984)

28. Steven Snobelen. 'Isaac Newton, heretic: the strategies of a Nicodemite.' *BJHS*, 1999, 32: 381–419

29. See Maurice Wiles, Archetypal Heresy (Oxford University Press, 1996).

30. Pre-existence of the independent thinking soul is a Socratic/Platonic discourse (see Plato's *Phaedo*, 76c-d cited by Gareth Matthews, 'Internalist Reasoning in Augustine for Mind-Body Dualism.' In J. P. Wright and P. Potter (Oxford Clarendon Press, 1999): pp. 133–4)

31. From Locke to Coleridge and Godwin, the lack of activity of mind during sleep was also commonly raised as proof of the interdependence of mind and body, whilst dreams were provided the counter thrust for independence of the soul.

32. Priestley, *Disquisitions* (1777): p. 305

33. *Free Discussion* (1778): pp. 70–71 and 80–81. David Hartley's beliefs were compatible with this Arian standpoint, as were those of Newton, his intellectual forebear.

34. *e.g.*, entry for 'Socinianism' in the glossary of Priestman, *M. Romantic atheism : poetry and freethought*, 1780–1830 (Cambridge University Press, 1999).

35. See also *A brief history of the Unitarians, called also Socinians* (1687): pp. 24–38. Recent theology has supported the notion. In response, critics also claimed that his particular philosophy had antecedents in antiquity, but in this case, likened his materialism to that of the Jewish Sadducees. e.g. see Anon *The Sadducee a Poem. Occasioned by Several Publications, and Particularly Disquisitions Relating to Matter and Spirit, by Joseph Priestley, L. L. D., F. R. S.* (London, 1778).

36. *Disquisitions* (1777): pp. 246–7

37. *Free Discussion* (1778): pp. 81–83

38. Excerpt from *Religious musings, a desultory poem*, written on the Christmas Eve of 1794 by Samuel Taylor Coleridge: 365–376) The "he" in lines 3–5 refers to David Hartley (cf. DNB entry for Hartley), whom Coleridge held in such great esteem during in his younger adulthood that he named his son after him.

Bibliography

Anon. [Stephen Nye] *A Brief History of the Unitarians, called also Socinians* (1687)

Anon *The Sadducee a Poem. Occasioned by Several Publications, and Particularly Disquisitions Relating to Matter and Spirit, by Joseph Priestley, L. L. D., F. R. S. (London, 1778).*

Coleridge, S. T. *The complete poetical works of Samuel Taylor Coleridge* (Oxford: Clarendon, 1912)

Dennett D. C. *Freedom Evolves* (Allen Lane, 2004)

Dictionary of National Biography (DNB) CD Rom. (OUP, 1995)

Golinski, J. *Science as Public Culture: Chemistry and Enlightenment in Britain, 1760–1820* (Cambridge University Press, 1999)

Hartley, D. *Observations on Man, His Frame, His Duty and His Expectations* 2 Vols. (1749)

Israel, J. *Radical Enlightenment: Philosophy and the Making of Modernity 1650–1750* (Oxford University Press, 2001)

Lawrence, W. *Lectures on Physiology, Zoology and the Natural History of Man* (London: 1819)

Locke, J. *Essay Concerning Human Understanding* (London, 1689)

Matthews, G. "Internalist Reasoning in Augustine for Mind-Body Dualism" in J. P. Wright and P. Potter (eds.) *Psyche and Soma: Physicians and Metaphysicians on the Mind-Body Problem from Antiquity to Enlightenment.* (Clarendon Press, Oxford; 2000)

Priestley, J. *Experiments and Observations on Different Kinds of Air* (1775)

– *Hartley's Theory of the Human Mind on the Principle of the Association of Ideas; with Essays relating to the Subject of it.* (London, 1775)

– *Disquisitions Relating to Matter and Spirit. To which is added, The History of the Philosophical Doctrine concerning the Origin of the Soul, and the Nature of Matter; with its Influence to the Doctrine of the Pre-existence of Christ* (London, 1777)

– *Doctrine of Philosophical Necessity an appendix to the Disquisitions relating to matter and spirit. To which is added an answer to the Letters on materialism and on Hartley's Theory of the mind.* (London, 1777)

– *A Free Discussion of the doctrines of Materialism and Philosophical Necessity, In correspondence between Dr. Price and Dr. Priestley. To which are added, by Dr. Priestley, an Introduction Explaining the Nature of the Controversy, and Letters to several Writers who have animadverted on his Disquisitions relating to matter and spirit, or his treatise on Necessity.* (J. Johnson and T. Cadell, 1778)

– *Memoirs of Dr. Joseph Priestley (Written by himself)* in *Autobiography of Joseph Priestley* (with Introduction by Lindsay, J) (Adams & Dart, Bath; 1970)

Priestman, M.*Romantic Atheism : Poetry and Freethought, 1780–1830* (Cambridge University Press, 1999)

Schaffer, S "'Priestley and the Politics of Spirit." in Anderson, R G W & Lawrence, C Eds. *Science, Medicine and Dissent: Joseph Priestley (1733–1804)* (Wellcome Trust/Science Museum, London 1987.): pp. 39–54

Smith, C. U. M. 'David Hartley's Newtonian Neuropsychology' *Journal of the History of the Behavioural Sciences 1987, 23, 123–136:*

Schofield, R. E. *The Enlightenment of Joseph Priestley : a Study of his Life and Work from 1733 to 1773* (Pennsylvania State University Press, 1997)

Snobelen, S. D. 'Isaac Newton, heretic: the strategies of a Nicodemite.' *British Journal for the History of Science*, 1999, 32: 381–419

Wiles, M. *Archetypal Heresy: Arianism Through the Ages* (Oxford University Press, 1996)

Yolton, J. W. *Thinking matter: materialism in eighteenth century Britain* (Blackwell, 1984)

THE FIRST ANAESTHETIC AND A TRIP IN A SUBMARINE*

Alan Dronsfield, Margaret Hill and Anna-Maria Rollin
University of Derby

Alan Dronsfield

The first reported physiological effect of oxygen dates back to 1775. Joseph Priestley announced "I have gratified that curiosity by breathing it ... The feelings in my lungs was not sensibly different from that of common air; but I fancied that my breast felt peculiarly light and easy for some time afterwards. Who can tell but that, in time, this pure air may become a fashionable article in luxury. Hitherto only two mice and myself have had the privilege of breathing it ... It may be conjectured that it might be peculiarly salutary to the lungs in certain morbid cases, when common air would not be sufficient to carry off the putrid effluvium fast enough. But though pure dephlogisticated air (oxygen) might be very useful as a medicine, it might not be so proper for us in the usual healthy state of the body; for, as a candle burns out much faster in the dephlogisticated than in common air, so we might, as may be said, live out too fast"[1].

Oxygen in early medical practice

Oxygen, arguably the most important of the 'hospital gases', is usually considered to have been discovered in the early 1770's. But it might have a history going back many years before this.

Priestley's prediction as to its likely usefulness to assist the lungs in pulmonary diseases was put into practice in 1783 when the French physician Caillens reported that he had successfully cured two cases of tuberculosis (phthisis) by means of oxygen therapy. That oxygen, along with other gases such as nitrous oxide, might have medical potential was perceived by Thomas Beddoes who established the Pneumatic Institute in 1798 for the treatment of "consumption, asthma, palsy, obstinate venereal diseases ... and other diseases which ordinary means have failed to remove"[2]. The bold claims of the Institute were not realised and moreover there were problems with impurities in the gases which made some patients worse rather than better. It closed down in 1802. Oxygen's use as a therapeutic agent continued sporadically through the nineteenth century. Some claimed its effectiveness against a diverse, but remarkably improbable, range of diseases: epilepsy, indigestion, tetanus, hysteria, 'fatty placenta', diabetes, leukaemia and many more.

For those patients who could not, or who chose not to, inhale the gas, other means of administration were available. In 1888, J. H. Kellogg advocated the use of oxygen enemas and cited their use in curing a patient of 'chronic lithiasis'. By 1890 subcutaneous injections were being tried. Curiously, neither

* Reproduced by kind permission of the Editor of Education in Chemistry.

method of administration appears to have caught on. Its first use in British anaesthetic practice dates from April 1847, when James Robinson, a dentist of Gower Street, London, used oxygen after ether anaesthesia "as a means of resuscitating patients after inhaling the vapour of ether"[3].

Thomas Beddoes

Lawrence Martin sums up the state of oxygen as a medical agent in the 150 years after its reputed discovery by Scheele or Priestley "The therapeutic use of oxygen was sporadic, erratic, controversial, beset with quackery and only occasionally helpful ... Oxygen therapy was characterised by methods which could not have resulted in much physiologic benefit"[2]. But the picture changed in the early years of the 20th century thanks to the rigorously scientific work by J. S Haldane, Joseph Barcroft and others and oxygen administration became routine to support anaesthesia and to ease breathing both in hospital and at home. Today an average size District General Hospital in the UK uses about 300 million dm^3 of gaseous oxygen per year. Typically two thirds of this is supplied in the liquid form and stored in giant insulated containers and the remainder is delivered in pressurised cylinders.

Primacy of discovery

Martin, along with most of the rest of the world, considers oxygen to be a discovery of the late 18th century. The general consensus is that it was discovered (but not reported) by Carl Wilhelm Scheele in 1771/2. He sent his findings for publication in 1775, but they did not appear in print until 1777, by which time Priestley had published "Experiments & Observations on Different Kinds of Air" (1775), the first chronicling of oxygen gas as a discrete species, different from atmospheric air.

However there are two schools of thought which ascribe the discovery of oxygen to workers in earlier times: the Sendivogius school, and the Drebbel school. In 1604 Michael Sendivogius (1566–1636) published an alchemical work on the Philosopher's Stone which ran to many editions, including an English one of 1650. He writes "Man was created of the Earth and lives by virtue of the Air. For there is in the Air a secret food of life, which in the night we call dew, and in the day, rarefied water, whose invisible congealed spirit is better and more precious than the whole universal earth". Sendivogius's supporters consider that this is the first time that air is revealed to be a mixture, and not a single substance according to the Aristotelian view of the world[4]. However the view that the air is more than a single substance predates Sendivogius. Certainly, Leonardo Da Vinci was writing around 1500 "flame is produced where a current of air is generated which serves to make the flame luminous. The fire constantly destroys the part of the air which

J. H. Kellogg

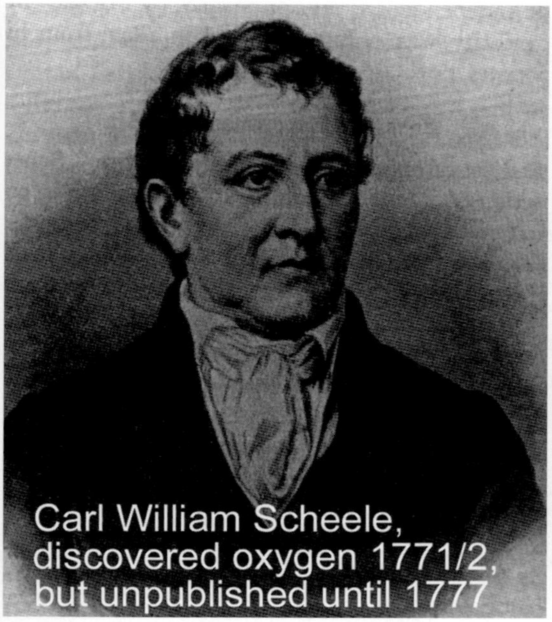

nourishes it and would form a vacuum if more air did not stream in to fill it"[5].

Sendivogius connected his aerial food to the salt potassium nitrate (nitre), the main source of which was farmers' dunghills: "Thus when it rains, the rain takes from the air a certain vital force and joins it to the nitrous salt of the earth ..." Alchemical historian Zbigniew Szydlo deduces that "Aerial nitre – what modern chemists call oxygen – seemed to be the key to nature; in gaseous form, it made all animal life possible; condensed into solid form as saltpetre (nitre) it gave life to plants and minerals"[4]. But there is no evidence that Sendivogius actually isolated his aerial nitre and, despite his interest in saltpetre, did not know that its thermal decomposition would yield oxygen. Interest in aerial nitre persisted long after Sendivogius's death (1636). Oxford chemist, John Mayow, was writing on saltpetre and aerial nitre in 1674. He reported, incorrectly, the action of heat on the salt, recording that it was decomposed into nitric acid and "alkali". Chemical historian James Partington considers that if he had actually tried the experiment, he might have isolated oxygen some 100 years before Scheele and Priestley[5].

Michael Sendivogius

Cornelis Drebbel (1572–1633) was an alchemist and inventor. Considered by some to be a charlatan, others such as Robert Boyle held his writings in high regard. The chemist and chemical historian Hermann Kopp says that his most useful discovery was the brightening effect that stannous chloride has on the red hues from cochineal and madder. What then is the connection between Drebbel and the discovery of oxygen? There are three parts to the answer. Firstly there is the statement which connects to the work of Sendivogius and to the later comments of Mayow: "Saltpetre is broken down by the power of fire, and so changed into the nature of air"[5]. Presumably this refers to the change (using today's formulae):

John Mayow

$$2NaNO_3 \rightarrow 2NaNO_2 + O_2$$

However nothing more is made of this report and sceptics may wonder if "the nature of air" can really be an allusion to oxygen. Secondly, elsewhere in the same publication he includes a wood-cut of a retort suspended by chains over a fire. The end of the neck dips into a bowl of liquid, presumably water, from which bubbles of gas are escaping. This would be indeed the apparatus to use to demonstrate the thermal decomposition of saltpetre and some authors are happy to make the connection[6]. But Drebbel's point about the experiment is that the fluid is drawn back into the retort on extinguishing the fire. Partington and others feel that had Drebbel used saltpetre in the retort he would have said so[5]. The illustration is completely disconnected from his work on nitre. The third supposition that Drebbel discovered oxygen comes from the fact, difficult to dispute, that he was one of the first men to construct a successful submarine and (by inference) he had some means of preventing his underwater oarsmen dying from suffocation. Drebbel's supporters conclude that only by having a supply of oxygen would this be possible.

The strange case of the submarine

The craft is reported as having travelled along the Thames from Westminster to Greenwich in 1620 (some say 1624, others 1615, but the latter is doubtful)[7], the journey taking three hours. There are no reports from witnesses of the event and we have to rely on second-hand descriptions appearing over the next 50 years.

Robert Boyle, one of the most noted chemists of the 17th century, was particularly intrigued by the means of maintaining a wholesome atmosphere for the crew. He made enquiries of Drebbel's son-in-law who described the use of a "chemical liquor which he accounted for the chief secret of his submarine *Navigation*. For when from time to time he perceived that the finer and pure part of air had been consumed, or over-clogged by respiration … he would, by unstopping a vessel full of this liquor, speedily restore to the troubled air such a proportion of vital parts, as would make again for a good while, fit for respiration"[8].

The nature of this liquor is lost in time. Apparently Drebbel communicated its identity to one man, who reported its action to Boyle. But Boyle, alas, says no more about it. There are two schools of thought as to what it might be.

One belief is that the "liquor" in the vessels was gaseous oxygen, and connecting it to his earlier observations on saltpetre, surmise that this was the source of the gas. Others consider that the liquor in the vessels was a solution of potassium hydroxide. This would serve to absorb the CO_2 produced by respiration, thus freshening the atmosphere, at least for a time. There are drawbacks with each theory. Numerous vessels of oxygen (at atmospheric pressure) would be needed to revitalise the air, and there is the problem of persuading it to leave the containers to do its job. With the potassium hydroxide theory, this certainly would not restore the oxygen content within the submarine and moreover, without any means of driving the air through the KOH solution, absorption would be too slow to be effective.

The time has come to apply 20th century science to Drebbel's submarine. Discounting a recent humorous suggestion that the vessel be constructed, and the voyage replicated by healthy (but possibly foolhardy) oarsmen[9], it is possible to test the feasibility of Drebbel's submarine by making some assumptions about its size, and then calculating how long the air contained in it would last. In the absence of contemporary diagrams of the craft we will assume:

- That it was powered by 12 oarsmen, rowing vigorously
- That there were no passengers to use up additional air
- That the voyage lasted three hours (if possible)
- That the dimensions of the craft were 7 m long, 2 m wide and 1.5 m high. This would be consistent with comfort, yet not too large to be burdensome for the rowers to propel.

Thus the total capacity of air would be 21 m^3, less the volume occupied by the actual bodies of the crew. Let us presume that the submariners had at their disposal 20 m^3 (= 20000 dm^3) of air, of which 20% (4000 dm^3) is oxygen.

A man at rest consumes 0.25 dm^3 of oxygen per minute, but under conditions of vigorous exercise, this rises to 2.5 dm^3 per minute. Our twelve vigorous rowers would consume 30 dm^3 per minute. Now serious problems arise when the oxygen level in the atmosphere falls to less than 10%. Using this as the threshold, the time taken to consume 2000 dm^3 of oxygen will be 67 min. Using the putative three hour journey from Westminster to Greenwich, our men would have expired around Wapping.

As the rowers consume their oxygen, they expire carbon dioxide. With a respiratory quotient of 0.8, consuming 2.5 dm^3 of oxygen per minute will give rise to an output of 2.0 dm^3 of CO_2 per minute. Our twelve rowers will collectively produce 24 dm^3 CO_2 per minute. Carbon dioxide narcosis leads to death within a few minutes when the volume concentration exceeds 17%. At exposures between 10 and 15% the symptoms are drowsiness, severe muscle twitching, dizziness and unconsciousness[10]. Let us assume a threshold value of 12% CO_2. This corresponds to a volume of 2400 dm^3 in the submarine. With a total production rate of 24 dm^3 per minute, this value will be exceeded after 100 min. Our vessel would have got as far as Limehouse.

What of Drebbel's method of revivifying the submarine atmosphere? Let us suppose his "fluid" was gaseous oxygen. We might be extraordinarily charitable and presume he had means to increase the oxygen content of the vessel's atmosphere to 50%. Vigorous rowing would reduce this to 10% as 8000 dm^3 of oxygen was consumed. This would have taken 267 min. The journey was theoretically possible. But the problem of CO_2 narcosis leading to death remains, and would be the same as before – after 100 min. Let us suppose Drebbel knew about caustic potash (potassium hydroxide) and its ability to absorb CO_2. This compound was known to the Romans and was prepared by "causticising wood ash" (presumably adding lime to impure potassium carbonate). The essential reaction with carbon dioxide is:

$$CO_2 + KOH \rightarrow KHCO_3$$

It is said to be rapid. Further reaction to form K_2CO_3 is possible, but this stage is slower. If we assume a successful three hour journey, our rowers will exhale 180×24 dm^3 of CO_2. Now 24 dm^3 of carbon dioxide has a mass of 44 g at room temperature and pressure, so the mass of CO_2 produced will be 7.92 kg. From the above equation, 44 g of CO_2 will react with 56 g of KOH. By proportion, 7.92 kg of CO_2 will react with 10.1 kg of KOH. Causticising wood ash affords only a very dilute solution of KOH, but let us assume that Drebbel could concentrate his liquor up to a 5% concentration. This would necessitate some 200 one litre flagons to be contained within the submarine, to be unstoppered as and when necessary. Readers may consider this to be unlikely. Furthermore, although the combination of CO_2 and KOH is rapid, we do not know how Drebbel brought the two species together for reaction. Certainly allowing diffusion alone to effect the chemistry would have been an impossibly slow process.

Regretfully, we come round to the conclusion that Drebbel's submarine voyage as described by legend, in the light of our calculations, is probably unlikely. Did he actually build a submarine? This may have had a foundation in fact, as notions about under-water travel can be traced back until 1580. But it is a pity that no first-hand descriptions of its construction survive. It is tempting to connect his 'voyage' with his discovery of oxygen, predating the works of Scheele and Priestley by some 150 years, but this, we fear is hardly supported by facts or theories. Today's attribution of the discovery of oxygen, so vital in hospital practice, is probably correct: Scheele first (but unpublished) and Priestley second (but published first)[11].

Acknowledgements

Much of today's knowledge of Michael Sendivogius comes from the researches of Dr. Zbigniew Szydlo[4,8]. His 1994 book is also a comprehensive source of information on the 17th century chemist John Mayow. We thank the editor of

Education in Chemistry and coauthors for permission to publish this presentation part of which originally appeared in 2004[12].

References

1. L. Martin, Oxygen therapy: the first 150 years, part 1 h accessed 24.10.01.
2. L. Martin . Oxygen therapy: the first 150 years, part 2 http://www.tsinai.org/pulmonary/papers/ox-hist/ox-hist1.html, Accessed 24.10.01.
3. J. Robinson, *Lancet*, 1847, **1**, 371.
4. Z. Szydlo, Hist. *Today*, 1997, **47**, 17.
5. J. R. Partington, *A History of Chemistry, Vol.* 2. London: Macmillan, 1961.
6. J. W. van Spronsen, *J Chem. Ed.* 1977; **54**: 157.
7. G. Tierie, *Cornelius Drebbel* (1572–1633). Amsterdam: HJ Paris Press, 1937.
8. R. Boyle, *New experiments. Physico-mechanicall, touching the spring of air, and its effects.* Oxford, 1660. Accessible in Z. Szydlo, Ambix, 1966, 43(2), 80 and Z. Szydlo, *Water which does not wet hands.* Warsaw: Polish Academy of Sciences, 1994.
9. A. T. Dronsfield and M. Hill, *Educ. Chem.* 2001, **38**, 140.
10. United States Environmental Protection Agency Report: http://www.epa.gov/ozone/title6/snap/co2report.html.
11. C. A. B. McLaren, *Proc. Hist. Anaesthesia.* 2000, **28**, 19 and A. J. Ihde, *The development of modern chemistry.* New York: Harper & Row, 1964. The issue of the primacy of oxygen's discovery has recently been addressed in a play: C. Djerassi and R. Hoffmann, *Oxygen.* New York: John Wiley and Sons, 2001.
12. A. T. Dronsfield, M. Hill and A. M. Rollin, *Educ. Chem.*, 2004, **41**, 68.

SECTION 12

BOC Symposium: Education and Religion

SCIENCE EDUCATION IN THE 18TH CENTURY

Norman Rose
University of Birmingham

Norman Rose

It is important to appreciate at the outset that, in the 18th century, some 80% of the population, in Britain, received no education whatsoever, other than what they might attain by their own efforts. The great bulk of the population was illiterate and innumerate. Monarchies, nobility, the landed gentry desired it that way, for education meant knowledge and information, which would clearly be dangerous in the hands of the great masses that constituted the lower and lowest social strata. A further 10% of the population acquired a very basic education, usually the rudiments of the three Rs, in order to pursue their trades and other occupations. This sector, however, was on the increase, especially in Britain, in response to the spur and demands of the burgeoning Industrial Revolution. Moreover, there was a growing awareness among the great illiterate mass, that only through education could they escape harsh penury and rise in social status.

Mainstream education, as proffered by the elementary feeder schools, grammar schools and the universities, in Britain had remained virtually unchanged for centuries. The ossified curriculum in place embraced divinity and the scriptures, Latin language and literature, Canon law, arithmetic and sometimes the rudiments of astronomy and music. At the tertiary level, the universities added Greek, Hebrew and the elements of medicine. Until well into the 18th century Latin remained the principal language within teaching establishments, whose primary reason for existence, both declared and actual, was to train students for the priesthood and other ancillary church duties. No concessions or modification to the curriculum were accorded to the increasing number who intended to follow other professions after graduation. All teachers and schoolmasters were Anglican priests for, by law, only subscribing and practising Anglicans could teach or learn at this time; mainstream education being under the complete domination of the church.

There were exceptions to all this. While Britain consisted of England, the whole of Ireland, Scotland and Wales, the Scots, for their part, had been particularly canny. Whilst accepting the political, defence and economic benefits of the Union of the Thistle and Rose in 1707, they had insisted on: retaining their Scottish law; to preserve their system of parish schools for the working and labouring classes as advocated by John Knox; and to retain Presbyterianism as the established church of Scotland. Although these conditions grated badly with the national government at Westminster and especially Anglican authorities in England, they were reluctantly accepted in order to secure the Union. Thus, in the 18th century, Scotland was altogether more liberal and enlightened than the rest of Britain. There was a degree of basic education for the masses and the five Scottish universities (Aberdeen had two) were far superior to the two in England with curricula that included law, economics, philosophy, engineering and science, but especially medicine.

Education was also more enlightened, if a little patchy, in the colonies of North America, almost certainly due to a substantial leavening of puritans, quakers and evangelicals among their numbers, who regarded education as both essential and everyone's right. Thus parish primary schools, along the Scottish pattern, were quickly established. On the near continent of Europe, education was a little more liberal and not completely under the dominance of church and state, as in most of Britain. Private, fee-paying schools, with a more modern curriculum than generally existed in Britain, flourished in France, the Netherlands, parts of Italy and in the states and principalities which eventually became Germany. But in all cases this education excluded the great bulk of the population, if only because they could not pay. In Britain, those that could not accept the thirty-nine articles of the Anglican faith and accompanying episcopacy, and who came to be called 'dissenters', began to establish their own schools or academies in defiance of the law. These laws, such as the Test and Corporation Acts enacted in the years following the Restoration of the monarchy in 1660, were particularly harsh and specifically designed to deter non-conformity of religion. Despite these laws many of the dissenting academies flourished, not least because they offered a far superior education open to students of any religious faith, or no belief at all. One of these, Warrington Academy, was for 30 years arguably the finest educational establishment in the world, largely due to the input and influence of Joseph Priestley. But that is a story which belongs elsewhere.

Earliest science education

There was no formal science education in the 18th century, with one minor exception to which I shall return later. That does not imply that science carried no interest or had no following. Some scientific research was undertaken by individuals at that time, and had been for centuries, all of which is part of recorded history. Scientific topics were also being studied by small groups in several countries including Britain, North America, and countries in Western Europe. But nowhere was this education formalized – and only in a few isolated instances did science form part of the curriculum. From the earliest times the topic was known as *natural philosophy* and this label applied throughout the 18th century and into the next one. It should be noted that there were no professional scientists in the 18th century or before that time. Aristotle and other philosophers of ancient Greece studied any subject they chose to, from astronomy to zoology, and because of their elevated status in society enjoyed a life of comparative affluence and luxury. From medieval times science was studied by gifted amateurs (dilettantes) such as Roger Bacon (13th c.), Copernicus (15/16th c.), Robert Boyle (17th c.) who either funded their own researches or were backed by wealthy patrons. A small number of individuals were fortunate enough – or astute enough – to gain a salaried sinecure, which gave them free rein to indulge in research: Isaac Newton, Robert Hooke, Antoine Lavoisier and Frederick William Herschel were among these. Most commonly, however, they earned their livelihoods in another capacity, such as schoolmaster, minister of religion or doctor of medicine, whilst pursuing their scientific interests as a hobby. It was by such channels that science advanced from the beginning until into the 19th century.

Science education where, for example, chemistry, physics, biology and botany were taught as part of the curriculum and a component of examinations, was virtually unknown. Not surprisingly, and in the absence of any convincing evidence, it can be concluded that science formed no part of primary education, anywhere. Nor did it play any significant rôle in secondary education. It is likely that, here and there, an individual schoolmaster or tutor in Britain or on the continent, because of his own self-interest in the topic, taught the rudiments of science, or one particular aspect of the subject, to his pupils, but they would have been isolated instances, and the evidence to support these is decidedly scanty. A definitive research might reveal that in the colonies of North America and then the United States, there were instances of science being part of secondary education.

There was some British science education at the tertiary level in the 18th century. But it was patchy, unstructured and without a common pattern. Even where it was taught and included in the curriculum, the subject was never part of the examination leading to a qualification or degree. A typical example of this is the position pertaining in Scotland. Most of the universities there taught some aspects of science which was popular among the student population. And student views, in Scotland, had to be respected because, unlike their English counterparts, the Scottish Universities were funded entirely from students' fees. Glasgow had a professor of anatomy and chemistry, and also one of mathematics. Edinburgh had a professor of natural philosophy (science) and also one of mathematics. Both universities had medicine faculties, which were considered among the finest in the world at that time, but these trained physicians and surgeons as part of a full academic study in which students sat examinations in order to qualify. So why the difference between medicine and other aspects of science? The answer is due to the immense popularity of science among thinking, educated people.

Science was generally confined to astronomy, mathematics and mechanics, which is not surprising in the wake of outstanding work in these areas, in the previous two centuries, by Brahe, Copernicus, Kepler, Leibniz and especially Newton who dominated all scientific thought and research in Britain and the continent of Europe throughout the 18th century. And so, in response to popular demand, universities taught these subjects but did not include them as a degree requirement. In the 18th century the works of Black, Lavoisier, Priestley and Scheele on chemistry, and Benjamin Franklin on physics took hold of the imagination so that these subjects also came to be taught, once again without leading to examinations.

The state of science education was roughly similar in the countries of Western Europe and North America, although the latter was somewhat ahead of the field, with some universities of the Eastern states offering chemistry degree courses by the end of the century. The two universities of England did not follow this general trend and it would be well into the 19th century before Oxford and Cambridge would begin to move away from their ossified curricula, which still retained many of the medieval trappings. But enthusiasm among the students in these universities to pursue scientific investigations was as great as anywhere else. Thus, in response to a growing demand, study groups began to form in some of the colleges, most often created and directed by tutors. Mathematics and physics became particularly popular at Cambridge, whilst at Oxford interest mainly centred on chemistry and biology. However, all these studies at both universities were unofficial, not forming part of the curricula and pursued as part-time leisure activities unrelated to the BA degree for which they had enrolled.

Scientific topics had always held great interest for the educated, and it was this that had given rise to the formation of such bodies as the Royal Societies of London and Edinburgh, and of their counterparts on the continent. In addition, groups of like-minded men had collaborated to create informal bodies such as the Lunar Society of Birmingham, formed in the late 1770s expressly to advance the cause of science and engineering. This group included such luminaries as Josiah Wedgwood, James Watt, Erasmus Darwin, Matthew Boulton and of course, Joseph Priestley. Benjamin Franklin, although living in England at the time, was never in the Lunar Society but followed its activities with great interest, and from time to time suggested individuals who should be included in the Society. As Professor Robert E. Schofield of America eloquently stated of this group:

> *These men were the harbingers of the Industrial Revolution. Together they comprised a clearing-house for the ideas which transformed their country materially, socially and culturally within a generation.*

There were other such groups that sprang up at that time. A 'College of Science applied to trades' was founded in Manchester in 1783; a similar body was brought together in Newcastle at the same time, and another formed in Derby a year later. None of these, however, made anything approaching the impact of the Lunar Society.

Earlier I stated that there was no formal science education in the 18th century, with one minor exception. There was a young man aged 25 years, newly appointed as Presbyterian minister to a small congregation in Nantwich, Cheshire. The duties were very light and the pay but £30 per annum, and so to fill his time and to improve his financial status he opened his own school. He immediately attracted thirty boys and six girls, at a fee of half-a-guinea per quarter. He instituted a very modern curriculum, compared with mainstream grammar schools, and this included mathematics, English language and literature, modern languages, the classics, history (English and European), some divinity and unbelievably, science! He was the first person, ever recorded, to teach laboratory science to schoolchildren. The material he taught was a conglomerate of general science and elementary physics. He encouraged his pupils to participate in the experiments and those in the highest class were charged with looking after the instruments and equipment and making use of them in their own time if they felt so motivated. Sometimes his scholars were called upon to lecture and demonstrate experiments to parents and friends in the evenings, which greatly enhanced the reputation of his school. His name – Joseph Priestley! Within a few years his growing renown as a teacher would take him 40 miles north as a tutor at the recently established Warrington Academy, where, within a short time, he introduced chemistry and physics to the curriculum.

Summing up, there was a great deal of interest in science in the 18th century and this led to scientific research by self-motivated individuals and groups often self-funded, or with patronage, but most usually as a part-time activity. But structured science education, entailing formal lessons and lectures as part of a curriculum was very rare, outside of the Scottish universities and Joseph Priestley's school at Nantwich. And science education leading to examinations as part of a degree was non-existent.

THEOLOGY UNDERLYING PRIESTLEY'S SCIENCE

C M P (Kate) Taylor
Yorkshire Unitarians and Open University

Kate Taylor

Priestley is known today as a man of science, notably in the fields of electricity and gases; he discovered oxygen – although he never fully understood what he had found – and, inter alia, he invented 'pop', or carbonated water. We talk about the influence of Benjamin Franklin and other of his fellow scientists, but we tend to ignore the influence of dissenters such as Theophilus Lindsey (1723–1808) or William Turner (1714–1794).[1]

Priestley was a true son of the Enlightenment; a wide range of fields was his playground and exploring them was a means to human progress. But Priestley was primarily a theologian and a non-conformist minister. Science was a leisure pursuit, a secondary, but by no means unrelated concern. In his *Memoirs* Priestley wrote: "I have always considered the office of a Christian minister as the most honourable upon earth and in the studies proper to it, I always took the greatest pleasure". Whilst there is much in his *Memoirs* about his scientific, and indeed other wider interests, the autobiography is, significantly, primarily about his path to Unitarianism and his work in his various ministries; among those friends and close acquaintances he mentions, other ministers are – if only just – foremost. My reason for speaking to you today is to flesh out Priestley's religious position and to underline its significance for his interest in and attitude to science.

Professor David Williams in his recent Essex Hall lecture, said that on studying Priestley, he became aware that "liberal views in religion might go hand in hand with a particular attitude to everything: a questioning of all received wisdom and a dissent from unsupported authority in any field of activity".[2]

Priestley was always a dissenter. His beliefs changed and developed, however. As a boy, living with his well-to-do and generous aunt, he attended a Calvinist independent chapel – the denomination known today as the United Reformed Church. Although he felt called to be a minister, Calvinism did not suit his temperament or intellect. He notes that "the remembrance of what I felt in that state of ignorance and darkness gives me a peculiar sense of the value of rational principles of religion".[3] His was an independent mind; he always wanted to think for himself.

Priestley was to reach his ultimate Unitarian position through discussion, close reading, and analysis. As a youth, at his aunt's house in Heckmondwike, he mixed with non-conformist ministers of varied views. Among them were men who were influenced by the 17th century Puritan minister and writer, Richard Baxter, who tried to soften Calvinism and preach salvation for all by faith rather than the strict Calvinist doctrine of predestination and salvation just for the elect. Calvinists catechised would-be members pretty strictly before admitting them. Baxter had wanted a more tolerant attitude. I find it curious, but probably significant, that the adjective Baxterian (and you can find it readily enough on the Internet) is not recorded in any of the modern, quality, dictionaries which lie about my living room.

Priestley's road to Unitarianism began there at Heckmondwike with his rejection of the doctrine of original sin. It debarred him from full membership of his aunt's church. He became an Arminian, rejecting the Calvinistic insistence on predestination.

As a non-conformist, Priestley was barred from taking a degree at either Oxford or Cambridge, and – as has no doubt been said many times in this symposium – he went instead, still bent in becoming a minister, to a dissenting academy at Daventry. His aunt had wanted him to go to a strictly Calvinist academy in London, Plasterers' Hall, but he rejected that since it would have meant subscribing to the ten articles of the Calvinist faith. What is important for us is that at Daventry Priestley was confirmed in his love of debate – of developing via discussion.

He went on, as a minister at Needham Market, to accept the Arian heresy that Jesus was divine but not equal to God the Father. He began to question the doctrine of atonement. The notion that Jesus died on the cross to save sinners was, he thought, untenable. During his years with Shelburne he became a materialist, accepting that the soul is not something separate from the body. Finally he became a Socinian – the name derives from Faustus Socinus – in believing that Jesus was simply a man, albeit divinely inspired and a pattern for human behaviour. There was no immaculate conception and no Virgin birth. The New Testament was a product of historical circumstances, not revelation. The Lord's Supper was not originally conceived of as a sacrament let alone as a mystery and only for the elect. At Mill Hill he had 'come out' as a Unitarian. He and Theophilus Lindsey, who founded the first avowedly Unitarian congregation, in London, in 1774, must be seen as the fathers of Unitarianism.

In his Memoirs he notes that his correspondence with Mr. Lindsey "has been the source of more real satisfaction than any other circumstance in my whole life".

Priestley did not, however, wish any brand of theology upon others. He was entirely happy for any minister and congregation to develop their own religious understanding – he was ready to see an increase in the number of sects rather than any decrease.

Discussion of, and written analysis of, the scriptures was a central Priestley activity. Investigation and reflection must lead, he thought, to improved knowledge. The idea of accepting an untested 'authority' would have been anathema. Unitarians today have a hymn which includes the phrase, "The Lord hath yet more light and truth to break forth from His Word".[4]

Seeking to disperse more light and truth, Priestley founded an irregular journal, *The Theological Repository*, the purpose of which was, as Priestley explained in his introduction to the first volume, "the chief good and happiness of man by advancing the moral perfection of his nature".

In his writings in the *Repository*, Priestley argues that the doctrine of the Virgin birth has no historical basis. Most importantly in terms of his theology he argues that "the point of the gospels is the promise of resurrection and a future state for all". The doctrine of a future state is the most striking thing in the Christian scheme. But he adds that the resurrection can only be relevant if we regard Jesus as a man like ourselves. It shows that bodily resurrection lies ahead for all.[5]

Where do Priestley's theology and his science come together? Well, simply, Priestley believed that God had "framed the world in such a way as to encourage man to seek truth through experimentation";[6] such experimentation – and it extended to the field of politics as well as science – would lead to greater, even perfect, human happiness Priestley believed, as all Unitarians still do today, in the importance of improvement in this world.

One of Priestley's methods – a favourite method I am tempted to say – of disseminating knowledge and understanding, was to gather and catalogue and, where possible, to try out or test … We can see this in his assembling a *History of the Corruptions of Christianity* (published only in 1782 but drafted much earlier). He aimed to get back to the core, pure, Christian knowledge by listing all the accretions which had been added and, indeed, accepted as doctrine. Priestley regarded this *History of Corruptions* as the most important of all his writings. There he argues that early writers regarded it as rather shameful to be the disciples of a man who died on a cross so they raised his status to that of God to make it more respectable! His *Institutes of Natural and Revealed Religion* is, again, a systematic statement of liberal dissent. His *History of Early Opinions Concerning Jesus Christ* (1786) is another catalogue intended to show how beliefs have developed and where error has occurred. You can see a similar, analysing and cataloguing approach in his *Biographical Chart*. And, of course, it is there in his *History and Present State of Electricity.*

Priestley had his intellectual limitations. Modern Unitarians regard it as quite strange that he should believe in the miracles of Jesus, especially when we consider that he was a determinist and that the principle of cause and effect is fundamental to science. And he believed in the resurrection of Jesus. Indeed he devotes a good deal of space in the *Theological Repository* to setting out its significance. It was not, as the more orthodox assert, an atonement but simply as a demonstration of what we can all hope for. Towards the end of his life he adopted a doctrine of universal restitution, of an after life in which moral progression can continue.[7] The limitations extended, of course, to his science; he never abandoned the Phlogiston theory, for example.

I will conclude by borrowing from Professor David Williams: "Priestley believed that science, religion and the arts were not separate activities; he took a holistic view of human activity and saw science as the exploration of the complexity of creation and therefore as a religious activity"[8].

NOTES AND REFERENCES

1. It was some writings of Turner that suggested the *Theological Repository* project to Priestley, See the entry on Turner in the *Dictionary of National Biography.*
2. David Williams, *Lessons from Joseph Priestley*, The Lindsey Press, 2004.
3. Joseph Priestley, *Memoirs*.
4. George Rawson, "We limit not the truth of God to man's poor reach of mind", in *Hymns for Living*, Lindsey Press, 1985.
5. Joseph Priestley, 'On the Life and Death of Christ', *Theological Repository*, Volume I, 1769.
6. Martin Fitzpatrick, 'Joseph Priestley', N*ational Unitarian Fellowship Viewpoint*, No. 173, February 2004.
7. See the article on Joseph Priestley in the *Dictionary of National Biography.*
8. David Williams, *op cit.*

Appendix I

REV. DR. JOSEPH PRIESTLEY – THE CHESHIRE CONNECTIONS 1758–67

Adapted for this publication from a Poster exhibited by Joe Lee
University of Manchester

Background

Joe Lee

Joseph Priestley was born at Fieldhead (now in Birstall) in 1733. After the death of his mother, Joseph at the age of 7 was sent to live with his aunt in Heckmondwike, Mrs. Sarah Keighley, a devout Calvinist; this set the seeds of his later dissenting religious beliefs (non-acceptance of Anglicanism). Because of persistent illness, his early education was sporadic, but during his period as a pupil at Batley Grammar School, he learned Latin, Greek, and (under the tuition of a Congregational clergyman) Hebrew and the rudiments of Chaldee, Syriac and Arabic. At the age of 19 in 1752, Priestley became a student at the Daventry Dissenting Academy for a period of 3 years, during which time his dissenting religious views developed through various forms of Unitarianism (non-acceptance of the Trinity). Between 1755 and 1758, he was a minister at a Presbyterian church in Needham Market, Suffolk with an annual salary of £30, which he supplemented from the establishment of a small school. This was an unhappy period; the congregation did not like his religious beliefs and he was treated coldly.

Sweet Briar Hall, Nantwich

Wall plaque on Sweet Briar Hall

Nantwich 1758–61

After Needham Market, Priestley transferred to a church in Nantwich where he lodged with a Mr. Eddowes, a grocer and senior member of his church. He found this appointment much more satisfying, particularly since the congregation accepted his theology. Here, again to supplement his income, he opened a small school for both boys and girls. Scientific interests began to develop and it was during this period that he bought an Air Pump, an Electricity Machine and some Optical Items, all of which were used in his teaching. For Priestley, science and religion were completely compatible. He was also taught to play the flute by Mr. Eddowes.

·61 ***Joseph Priestley*** **was the Unitarian minister at the chapel built by the Presbyterians in *1726*, on the corner of Hospital Street and Pratchett's Row.**

"At Nantwich I found a good natured friendly people with whom I lived happily for three years".

Memoirs of Rev.Dr. Joseph Priestley

Priestley established a successful school of thirty boys and six young ladies, equipping it with a small air pump and electrical machine so that they might perform scientific experiments. After he left *Nantwich* he was to become world-famous as the discoverer of oxygen and father of modern chemistry.

Part of Poster in Nantwich Museum

The Warrington Academy

*Warrington** *1761–67*

In 1761, Priestley was appointed Language Tutor at Warrington Dissenting Academy (a post he was to retain until 1767 when he was invited to become a minister at Mill Hill Chapel in Leeds). Much happened during the Warrington period:

1. In 1762, was ordained
2. In the same year, married Mary Wilkinson (aged 20, sister of one his Nantwich students and daughter of ironmaster Isaac Wilkinson, a leading figure in the Industrial Revolution), his wife for 34 years until her death.
3. Made a Doctor of Laws by the University of Edinburgh in 1765 and a Fellow of the Royal Society in 1767.
4. Publications included *The Rudiments of English Grammar* (1761), *A Chart of Biography* (1765), *Liberal Education for Civil and Active Life* (1765) and (encouraged by Benjamin Franklin) *The History and Present State of Electricity* (1767). Contents of the last of these clearly indicated the importance which Priestley attached to appreciating its history when studying a subject.
5. Member of the Committee for the Warrington Circulation Library.

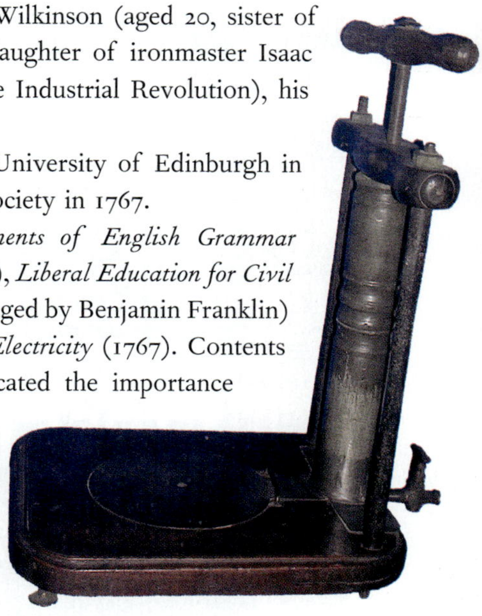

(*right*) Air pump

* Until 1974, Warrington was in the County of Lancashire, but is now in Cheshire.

Priestley's Warrington house

Plaque at Original Warrington Academy site

English Grammar book

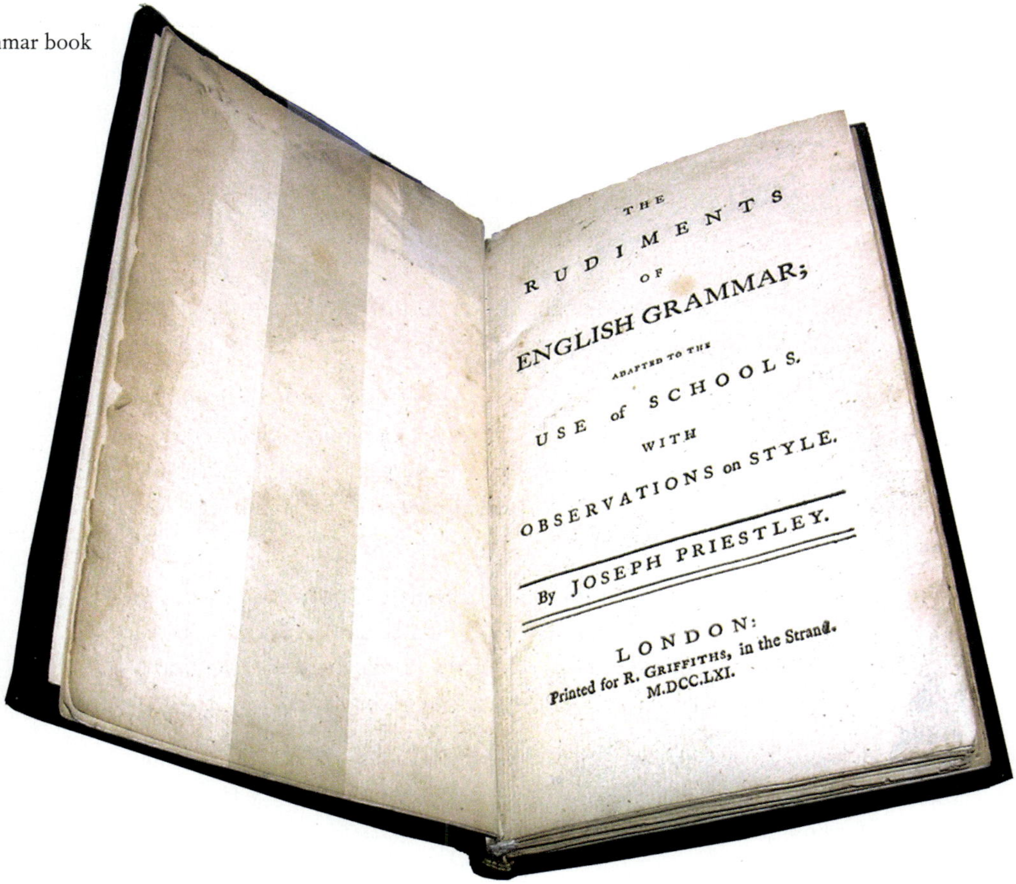

(*right*) Proceedings of Warrington Circulating Library. Revd J Priestley's name can be seen on the left page

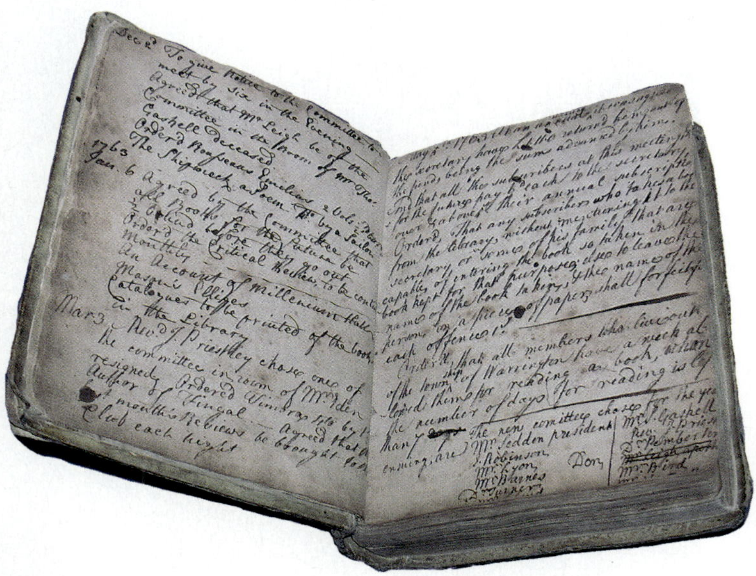

(*below*) Electricity book

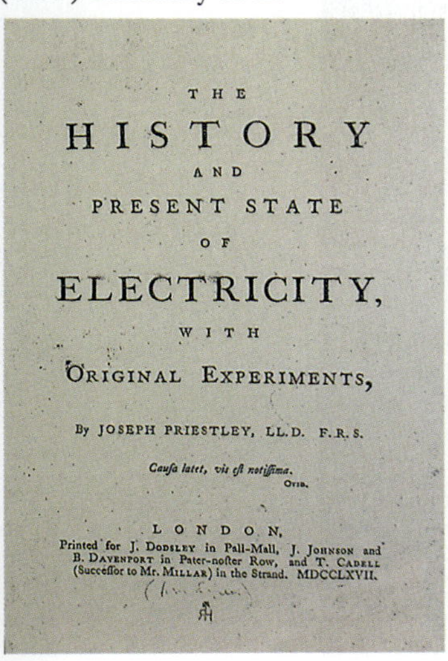

THE
HISTORY
AND
PRESENT STATE
OF
ELECTRICITY,
WITH
ORIGINAL EXPERIMENTS,

By JOSEPH PRIESTLEY, LL.D. F.R.S.

Causa latet, vis est notissima.
OVID.

LONDON,
Printed for J. DODSLEY in Pall-Mall, J. JOHNSON and B. DAVENPORT in Pater-noster Row, and T. CADELL (Successor to Mr. MILLAR) in the Strand. MDCCLXVII.

Appendix II

PRIESTLEY'S TEACHING GLOBES AT MANCHESTER COLLEGE, OXFORD

The celestial globe has no maker's name, but the terrestrial globe has a printed inscription (in a blank space in the Pacific Ocean), as follows:

> A new & correct globe of ye Earth, together with a view of ye General and Coasting Trade Winds, Monsoons &c Laid down according to ye Newest Discoveries & from ye most Exact Observations By John Senex F. R. S. Now made and sold with very Considerable Improvements by B. Martin of Fleet Street.

There is no date on the globes, but they clearly predate Captain Cook's voyage which surveyed the coastline of Australia and New Zealand. Senex died in 1740 and Martin died in 1782. On the rim of each of the stands there are two kidney-shaped brass plates. The first plate on the celestial globe says:

> These globes belonged to the Revd. Joseph Priestley LL. D., F. R. S. the discoverer of oxygen 1774 they were given to the Revd. Pendlebury Houghton by one of whose descendents this tablet was added 1902.

The first plate on the terrestrial globe says:

> These globes narrowly escaped destruction when Dr. Priestley's library was burned by a Birmingham mob

The second tablet on both globes says:

> Presented to Manchester College Oxford by Mr. and Misses Enfield of Nottingham April 3rd 1923

Pendlebury Houghton was a student of Warrington Academy 1773–1778, Assistant Classical Tutor 1778–9, Minister at Dob Lane Manchester 1779–81, Shrewsbury 1781–87, Norwich 1787–1808 and 1810–12, Princess Street Westminster 1808–1810, Liverpool Paradise Street, 1812–23, died 1824. William Enfield was his (elder) colleague at Warrington Academy and at Norwich; no doubt the Enfields of Nottingham were his descendents. Houghton's death mask is now kept at Manchester College, Oxford. He was a very devotional preacher of a saintly character (according to his friends and admirers) and empty, rhetorical and namby-pamby (according to his enemies). At the time if his death, he had just been appointed Tutor in Rhetoric at Manchester College. He belonged in preaching to the school of Hugh Blair, the admired Scottish preacher.

There is at the College a pamphlet with the Globes, according to which "the globes had been given away to Houghton, a young and promising student, by Priestley long before the Birmingham riots". The brass plate inscription is therefore incorrect. In any case, Cook's voyage had made them out-of-date in reference to the Antipodes, so perhaps Houghton himself passed them on to the Enfield family (William Enfield died at Norwich in 1797, leaving two sons and three daughters), as ornamental rather than useful.

Priestley is said to have used the Globes at his school in Nantwich between 1758 and 1761.

Appendix III

PRIESTLEY COMMEMORATED IN COINS AND MEDALLIONS

Based upon a poster presentation by Roy A Olofson and Robert D Minard of Penn State University which was first exhibited at the Oakwell Hall Priestley Bicentennial Colloquium on February 6–7th 2004

Hancock Medal (3.5 cm diameter)

This medal by J. G. Hancock of Birmingham was struck in silver, bronze and tin, and possibly also gold in 1783 the year of Joseph Priestley's 50th birthday. The obverse has Priestley's profile with the inscription "Josephus Priestley". The reverse celebrates his accomplishments as a scientist by depicting some of his instruments of scientific research including the scientific trough, critical to several of his discoveries. With the aid of this apparatus, Priestley discovered many important gases including oxygen ('dephlogisticated air'), hydrogen chloride, ammonia, (alkaline air) sulfur dioxide (vitriolic air), hydrogen sulfide, three oxides of nitrogen, including the first surgical anaesthetic (laughing gas), silicon tetrafluoride and carbon monoxide. The last was made and characterized in Northumberland Pennsylvania, his final home. There are at least two versions of this medal. In the very fine silver version depicted on the right and centre the small furnace belching smoke is on the left and the gear system and electrical machine are on the right. In the more pested tin medal on the right, these are reversed and some of the instruments in the cabinet are different.

J. PRIESTLEY CITIZEN OF THE WORLD (2.9 cm diameter copper)

This political medal, minted anonymously ca. 1790 greatly exacerbated the antagonism of the Church of England and the Monarchy toward Priestley. While the bust is a copy of the Hancock medal, there are minor differences and this medal is smaller. The reverse features the flame of liberty proceeding from the tomb of "HAMPDEN & SIDNEY" and is topped by the radiating 'cap of liberty' and the caption "SACRED TO LIBERTY". The die flaw evident above is also in the photo in D. & H. Warks, the bible of British conder tokens.

Lunar Society Medal (5.1 cm diameter)

This medal, struck in Britannia silver in 1966 by the Birmingham mint, marks the bicentenary of the Lunar Society. Founded about 1766 this association of scientific men flourished in Birmingham for nearly 40 years. They met at each other's house for dinner at 2 pm every month on the Monday nearest to the full moon ("in order to have the benefits of the light in returning home") to exchange ideas in science and the humanities. Priestley wrote: "we had nothing to do with the religious or political principles of each other, we were united by a common love of science ...". Besides Priestley (a later arrival) members included James Watt (steam engine), industrialist Matthew Boulton (a partner of Watt, maker of Staffordshire silver plate, Boulton mint, the "Father of Birmingham"), Erasmus Darwin (physician, botanist, poet and grandfather of Charles Darwin), William Small (mathematician and founder member with Boulton and Darwin), James Keir (industrial chemist), William Withering (physician, foxglove, – digitalis – in medicine), Josiah Wedgwood (industrial potter, inventor), John Whitehurst (instrument maker), Richard Edgeworth (educationalist), Samuel Galton (gunsmith) and Thomas Day (social reformer). Periodic guests included Benjamin Franklin, Sir William Hershel (astronomer) and John Wilkinson (ironmaker and Priestley's brother-in-law) The other side of the medal has a stylized "LS", an eye, and the motto "OBSERVARE".

Priestley Medal of the American Chemical Society (7.6 cm diameter)

This large medal cast in 10 carat gold (obverse left) with a duplicate in bronze (reverse right) is the highest award of the American Chemical Society (ACS). It is presented annually "to recognize distinguished services to chemistry". Winners in the past 10 years include Nobel Laureates Sir D. H. R. Barton (British) and E. J. Corey, along with the 2003 $300000 Welch awardee Richard Breslow. The exact duplicates pictured above were presented to Joseph Priestley House Museum in Northumberland Pennsylvania USA for permanent display in 2001 as part of the 125th anniversary celebration of the ACS and of the role of Joseph Priestley House in the founding of the ACS at a dinner in the honour of Priestley on the first anniversary of oxygen.

Phipson Medal (5.4 cm diameter)

This was minted in 1794 in silver, tin and bronze by William Phipson, to commemorate Priestley's emigration to America in that year (departed Britain April 8th arrived June 4th). A replica of this medal struck in bronze in 1776 (by BASF as part of a series) is pictured above. The second 1804 edition of this medal (see above in tin) includes the dates of Priestley's birth and death. (Natus 13 Mar 1733 Mort 6 Feb 1804). Instead of "MAGNUS CHRISTIANUS PHILOSOPHUS", another version of this medal in bronze (in Leeds City museum) has the inscription "I DECUS I NOSTRUM MELLIORIBUS UTERE FATIS" (from Virgil's Aeneid VI 546)

Pandora's Breeches (2.8 cm diameter copper):

This medal by Spence Skidmore (1792) depicts a straw filled pair of breeches on fire with smoke billowing up from the beltline. Beneath lies a serpent with a sting for a tail and with the head of Joseph Priestley severed from its body with a dagger. An attempt to set the House of Commons on fire in this way (in May 1792) was discovered before the wooden laths caught fire and Parliament and the adjacent timber buildings destroyed. Priestley was rumoured by malevolent Church and King propagandists to be the inspirer of the plot and the reference is to the liberation of "evil" when Pandora's box was opened. In the medal (image) on the left, the reverse with its crown, anchor and hearts is irrelevant. More inflammatory medals with the same obverse shown to the right, have Thomas Paine hanging from a gallows on the reverse with the legend "END OF PAIN". Coupling Paine and Priestley may refer to both having been invited as members of the French national Convention in 1792. Priestley declined the invitation but the more radical Paine accepted.

Franklin Mint Oxygen Bicentennial Medal (3.8 cm diameter)

To commemorate the 200th Anniversary of Priestley's discovery of Oxygen on August 1st 1774, this medal was commissioned in both silver and bronze from the Franklin Mint by the American Chemical Society. It also celebrates the centennial of the first national Chemical Congress held in the United States (July 31st–Aug 1st 1874 in Northumberland Pennsylvania) in honour of Priestley, and led to the founding of the American Chemical Society two years later.

The Wedgwood Medallion

A 10 inch black plaster relief of Priestley in the Wedgwood Museum, modelled by Giuseppi Ceracci in 1778, was recreated as a medallion the same year. It was modified by William Hackwood (Wedgwood's chief modeller) in 1779 and also in white relief on the signature Wedgwood blue jasper background in several sizes. Editions from that period through to the late 20th century have been sold either separately or as part of other pieces. In this medallion made in 1952, Priestley's portrait is 2 inches tall.

Halliday Medal (5.1 cm diameter):

This was created in Bronze (pictured here) and in tin by Thomas Halliday in 1804 to mark Priestley's death. The reverse carries the inscription on the memorial tablet erected at the New Meeting Birmingham (now at Five Ways Edgbaston). This tablet is consecrated to the memory of Rev. JOSEPH PRIESTLEY LL. D. by his affectionate CONGREGATION in testimony of their GRATITUDE for his faithful attention to their Spititual Improvement ... His Discoveries as a Philosopher will never cease to be remembered and admired by the ablest improvers of Science. His Firmness as an advocate of Liberty and his Sincerity as an Expounder of the Scriptures endeared him to many of his enlightened and unprejudiced of his Contemporaries. His example as a Christian will be instructive to the Wise and interesting to the Good of every Country and in every Age ..." He was the Minister at New Meeting from Dec 31st 1780 till its destruction along with Priestley's home in the Birmingham riots on Bastille day in 1791.

New Meeting House Burned in 1791 Medal (2.9 cm in copper)

This medal commemorates the burning of Priestley's Chapel. The reverse of this identifies it as part of a series of coins depicting Birmingham's buildings by Peter Kempson.

End of Bunepart (sic) and the French Army (2.8 cm diameter copper)

Besides the slogan, the reverse side of this medal depicts a row of fully loaded gibbets extending to the far distance. Joseph Priestley, that despised "fellow traveller" of the French revolutionary movement, hangs from the first gallows. The obverse records the presentation of colours to the Birmingham Association. Napoleon had assembled an "Army of England" and during the invasion scare voluntary militias were assembled throughout England including a newly inaugurated cavalry company on Birmingham Heath on June 4th 1798.

Modern Commercial Medals (3.8 cm diameter)

These medals were struck in the 1970's and 1980's as parts of series including other famous scientists.

John Wilkinson, Iron Master (2.8 cm diameter)

Priestley's brother-in-law, and patron of the family, was the iron master John Wilkinson, whose foundry provided the iron for the great iron bridge over the Severn, and who built the first iron ship. The family business is now more famous for Wilkinson Sword razor blades. Wilkinson's high standard of boring cylinders made possible James Watt's rotary steam engine and this methodology also afforded better iron cannons. Because it was unprofitable for the monarchy, small coinage was minted under commercial franchise, and Wilkinson was one of the two largest manufacturers of halfpennies. The coin depicted above is a bronzed copper halfpenny. All the conder halfpennies have "JOHN WILKINSON IRON MASTER" and Wilkinson bust on the obverse, though the scene on the reverse varies. It is said that in each new edition, Wilkinson looks more like King George III.

Appendix IV

PRIESTLEY FAMILY TREE RESEARCH

Margaret Ware-Priestley and Jim Birch

Present at the Huddersfield meeting were Dr Priestley Toulmin, who gave the first annual July 4th Priestley Lecture, and Mrs Margaret Elizabeth Ware. They are, respectively, direct descendants of Joseph Priestley and his brother, Joshua. Both descendants have made thorough studies of their family history.

The results of their extensive researches are presented here as family trees.

The authentification for Dr. Toulmin's tree can be found in the following article:

P. Toulmin, "The Descendants of Joseph Priestley. LL. D., F. R. S. (a progress report)", Northumberland County Historical Society, Proceedings Vol 32, Pt 2, Pennsylvania 1994, 129 pp.

That for Mrs Ware's tree is covered by her following account.

Jim Birch

Margaret Priestley-Ware and Priestley Toulmin

> Father had always told us that Dr. Joseph Priestley was "some kind of uncle". The name was dying out, so to keep the connection alive for new generations, I decided to PROVE IT!

Simple matter! Work forwards from Dr Joseph's brothers and backwards from our grandparents.

I asked Aunt Phyllis (always go to the old ladies), who confirmed the name of great-grandfather Timothy. From Timothy I worked back to great-great-great-grandfather James. Later I discovered the family beyond Jonas, Dr. Joseph's father, in Hartshead.

Thirteen years later (some simple matter!) I was well and truly stuck, although by then I knew our ancestor *must* be Joshua – but where were his children baptised?

By coincidence – or was it good luck? – I was put in touch with another descendant of Joshua and everything fell into place. Only later in 2003 was I able to complete this Family Tree. Any errors or omissions are entirely mine.

Warm thanks to Huddersfield Public Libraries for use of micro-films/fiches of Census Returns 1841–91, Parish Registers for Birstall and Hartshead, International Genealogical Index and General Register Office, also *The Autobiography of Dr. Joseph Priestley* ed. Jack Lindsay, pub. Adams & Dart, Bath; Dewsbury Registry Office.

Thanks also to:

(a) Fellow researchers/genealogists John Scatchard, Geoffrey Keighley, John Priestley (no relation) Naomi Jarvis (N. Z.) and Andrea Bashore of Joseph Priestley House, U. S. A.

(b) Finally my sister Ann, her children Rachel and Will, cousin Stuart Lister and my long-suffering husband Derrick for their support and encouragement.

Three trees are presented on the following pages in the order: (a) The Toulmin tree, (b) The Ware tree, (c) A minimal direct lineage tree showing the relationship between Priestley Toulmin, Margaret Ware and Joseph Priestley.

The family trees produced on the following two pages were produced by the Family Historian package

The Toulmin Tree

Note that this tree also includes Priestley Toulmin's children and grandchild.

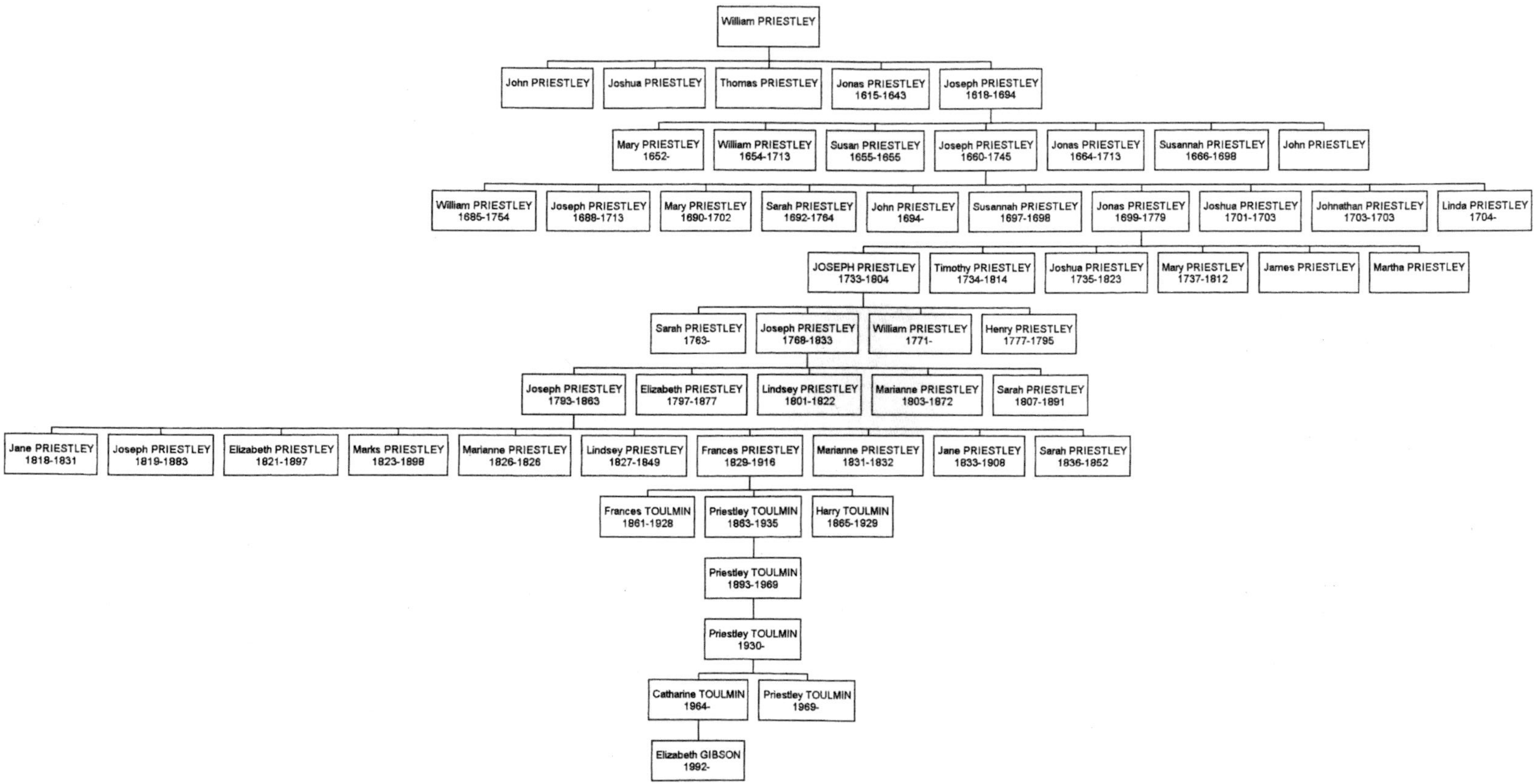

The Ware Tree

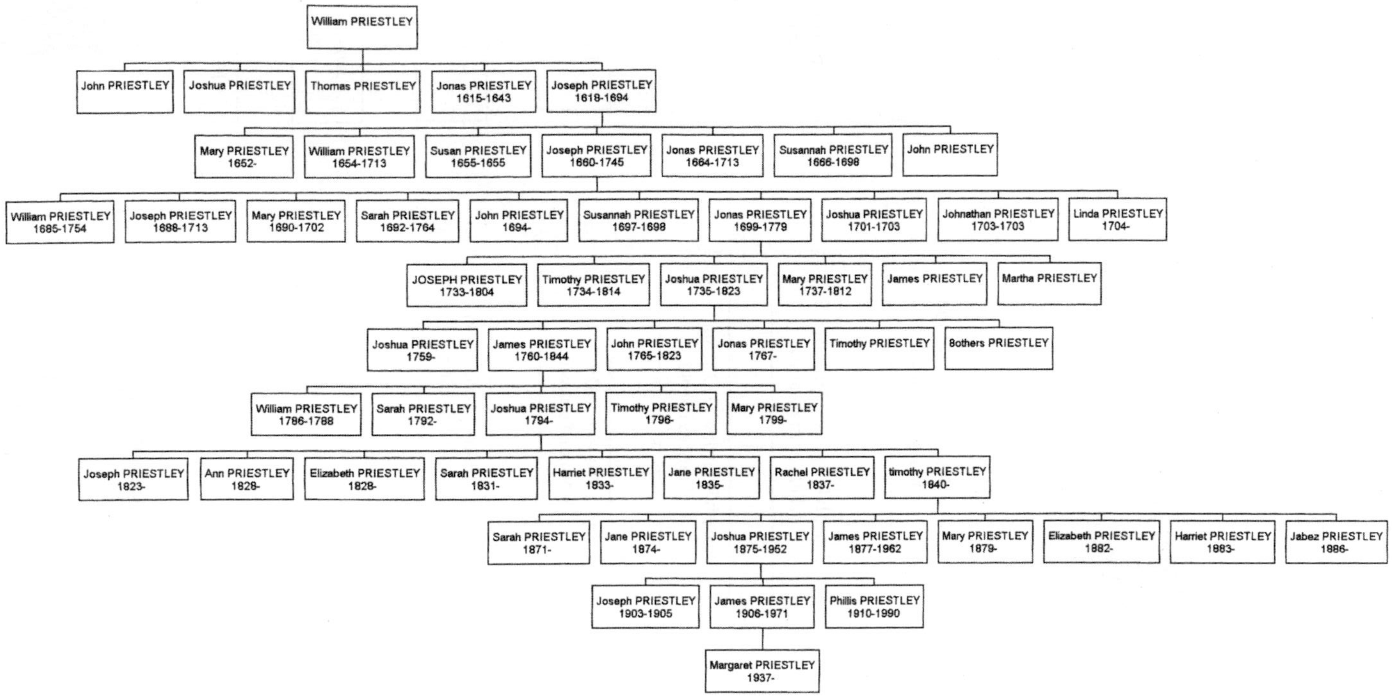

William Priestley

- Joseph Priestley 1618–1694 m. Jennet Taylor
  - Joseph Priestley 1660–1745 m. Sara Healey
    - Jonas Priestley 1699–1779 m. Mary Swift

Joseph Priestley 1733-1804 m. Mary Wilkinson

- Joseph Priestley 1768–1823 m. Elizabeth Rylands
- Joseph Priestley 1793–1863 m. Frances Biddle
- Frances Priestley 1829–1916 m. Major **Toulmin**
- Priestley Toulmin 1863–1935 m. Grace Chewning
- Priestley Toulmin 1893–1969 m. Catharine Carey
- **Priestley Toulmin 1930– m. Mary Slason**

Joshua Priestley 1735–1823 m. Mary Drake

- James Priestley 1760–1844 m. Rachel Fisher
- Joshua Priestley 1794–? m. Harriet Clough
- Timothy Priestley 1840–? m. Sarah Baxter
- Joshua Priestley 1875–1962 m. Annie Taft
- James Priestley 1906–1971 m. Maggie Schofield
- **Margaret Priestley 1937– m. Derrick Ware**

Appendix V

A SELECTION OF QUOTATIONS BY AND ABOUT JOSEPH PRIESTLEY

Compiled by Derek A. Davenport
Purdue University

Preface

Derek A.Davenport

It is surely unreasonable to ask an author whose theological works alone run to more than twenty large volumes that he be a master of style if not of concision. Priestley's writings are often infuriatingly prolix and repetitive and, while this can become a kind of virtue in the transparent honesty of *Experiment and Observations on Different Kinds of Air*, it does put an uncommon burden on the casual reader's patience. Priestley tells of writing at the fireside with his children about him and, like Shakespeare (in Ben Jonson's eyes), it sometimes seems as if "he never blotted out a line". One can only echo Jonson's "would he have blotted a thousand". But for Priestley, language was a tool and a fine weapon and he freely admitted, "my object was not to acquire the character of a fine writer, but of a useful one". When the occasion arose and the spirit moved Priestley could, however, be pithy or poetic, sage or savage, occasionally aphoristic through rarely witty. It is hoped that the following by no means random selection of quotations by and about Joseph Priestley may enable the reader to arrive at a plausible, but inevitable partial, portrait of this quite extraordinary man.

Priestley as Seen by Himself

If my writings in general have been useful to my contemporaries, I hope that this account of myself will not be without its use to those who may come after me, and especially in promoting virtue and piety, which, I hope I may say, it has been my care to practice myself, as it has been my business to inculcate them upon others.

Autobiography

My father, Jonas Priestley, was the youngest son of Joseph Priestley, a maker and dresser of woollen cloth. His first wife, my mother, was the only child of Joseph Swift, a farmer at Shafton, a village about six miles south – east of Wakefield. By this wife he had six children, four sons and two daughters. I, the oldest, was born on the 13th March, old style, 1733, my father married again in 1745, and by his second wife had three daughters.

Autobiography

From time to time I discovered any fondness for books, my aunt entertained hopes of my being a minister, and I readily entered into her views. But my ill health obliged me to turn my thoughts another way, and, with a view to trade, I learned the modern languages, French, Italian and High Dutch, without a master; and in the first and last of them I translated and wrote letters, for an uncle of mine who was a merchant, and who intended to put me into a counting-house in Lisbon. A house was actually engaged to receive me there, and everything was nearly ready for my undertaking the voyage. But getting better health, my former destination for the ministry was resumed ...

Autobiography

So they [his tutors] indulged us in the greatest freedoms, so that our lectures had often the air of friendly conversations on the subjects to which they related. We were permitted to ask whatever questions, and make whatever remarks we pleased; and we did it with the greatest, but without any offensive, freedom.

JRC, p. 53

All the time I was at the academy [Daventry} I never lost sight of the great object of my studies, which was the duties of a Christian minister, and there it was I laid the plan I have executed since.

JRC, p. 54*

Like most other men of liberal education, I had conceived a great aversion to the business of school master and said I would have recourse to anything else for maintenance in preference to it.

Autobiography

Few are qualified to make new discoveries – but when the discoveries have been made ... people of inferior abilities are sufficient to digest those principles into convenient method.

The History and Present State of Electricity with Original Experiments

A person who means to serve the cause of science effectually, must hazard his own reputation so far as to risk even *mistakes* in things of less moment. Among a multiplicity of new objects, and new relations, some will necessarily pass without sufficient attention; but if a man be not mistaken in the principal object of his pursuits, he has no occasion to distress himself about lesser things.

Experiments and Observations on Different Kinds of Air

He who does not foolishly affect to be above the failings of humanity, will not be mortified when it is proved that he is but a man.

Experiments and Observations on Different Kinds of Air

* See selected reference [6]

In general, during the composition of those works, the greatest part of every day was spent in my laboratory, and in the evenings and mornings only in reading or writing. Besides, these different studies so relieve one another, that I believe I do more in each of them, by applying to them alternately, than I should do, if I gave my whole attention to one of them only.

Experiments and Observations on Different Kinds of Air

You have destroyed the most truly valuable and useful apparatus of philosophical instruments that perhaps any individual, in this or any other country, was ever possessed of, in my use of which I annually spent large sums with no pecuniary view whatever, but only for the advancement of Science, for the benefit of my country and mankind. You have destroyed the Library corresponding to that apparatus, which no money can repurchase except in the course of time. But what I feel far more, you have destroyed manuscripts which have been the result of the laborious study of many years, and which I shall never be able to recompose; and this has been done to one who never did, or imagined, you any harm. In this business we are the sheep and you are the wolves. We will preserve our character and hope you will change yours. At all events we return you blessings for curses, and hope that you shall soon return to that industry and those sober manners for which the inhabitants of Birmingham were formerly distinguished.

J. Priestley to the Inhabitants of the town of Birmingham, July 19th 1791

I should have read my books to little purpose if they had not taught me to bear the loss of them with composure and resignation

(1791)

The country was in a better state to invite your return to it, but it is far otherwise, at least with respect to myself and those who have generally passed for the friends of liberty. Such is the spirit of bigotry encouraged by the Court party, that great numbers are going to America, and among others all my sons, and my intention is, that when they are settled to follow them and end my days there.

Letter to William Withering (1793)

I take the liberty to introduce to you Mr. Jones, who was lecturer in chemistry at the New College in Hackney, in which employment I now succeed him, and who is to be my successor at Birmingham. You will find him to be equally modest and sensible, and, as a philosopher, more inclined, I believe, to your system than to mine; but open, as we ought all to be, to convictions as new facts present themselves to us.

Only surviving letter to Antoine Lavoisier (1792)

I do not pretend to leave this country, where I have lived for so long, and so happily, without regret, but I consider it as necessary, and I hope that the same good providence that has attended me hitherto will attend me still.

Autobiography (1794)

I read the whole of the Greek Testament, and the Hebrew bible as far as the first Book of Samuel: also Ovid's Metamorphoses, Buchanan's poems, Erasmus' Dialogues, also Peter Pindar's poems, &c ... and to amuse myself I tried the heat of the water at different depths, and made other observations, which suggest various experiments, which I shall prosecute whenever I get my apparatus at liberty.

(1794)

It is no more than justice to say of these addresses – in the lump – that they are distinguished for a certain bareness of thought and vulgarity of stile, which, were we not in possession of the Doctor's answers, might be thought inimitable.

Peter Porcupine (William Cobbett) (1794)

There is something so pathetic, so irresistible moving in all this that a man must have a hard heart indeed to read it, and not burst into laughter.

Peter Porcupine (William Cobbett) (1794)

If he did not foresee them (objections), he must have an understanding little superior to an idiot; if he did he must have the heart of a Marat. Let him choose.

Peter Porcupine (William Cobbett) on Joseph Priestley's pamplet Marat, St Paul and Jesus Christ (1796)

For though I have made discoveries in some branches of chemistry, I never gave much attention to the common routine of it and know but little of its common processes.

Letter to University of Philadelphia (1794)

But of what importance is it where I was born, or whence I came ... Here I am. Here is my family. Here is my property, and everything else that can attach a man to any place. Let any person only view my house, my garden, my library, my laboratory, and the other conveniences, with which I am surrounded, and let him withal consider my age, and the little disposition that I have shewn to ramble any whither, and say whether any person among yourselves, or in the United States, could remove with more difficulty, or with more loss, than I should do.

Letter to England (1795)

I examined what remain'd of the flour, and cannot say that I found any appearance of Arsenic in it. Since therefore no real mischief has been done, and it cannot be proved that any was intended, I shall not make any further enquiry into the business …

I am as ever, your affectionate father …

William Priestley had been accused by the Federalist press of poisoning his father (1799)

I thank you for your hint of politics. How different is our situation from yours! Our debt is trifling, and will to appearance soon be discharged, though almost all our taxes are done away. Our particular state has a fund, out of which all the expenses of government are defrayed; so that we hear of nothing but county taxes for the repair of roads, &c. we in this place have only been called upon to pay a poor's rate, and in both cases on account of single insane persons, and we hardly ever see a beggar, … everybody contributing with great cheerfulness. There being no church establishment, we have no tithe … I do not think that any country in the world was ever in a state of greater improvement, in all respects, or had fairer prospects … It is a great pleasure to look round and see this: and yet we have many grumblers and discontented persons.

Letter to England (1802)

I now, for the first time in my life (and I shall soon enter my 70th year), find myself in any degree of favour with the government of the country in which I have lived, and I hope I shall die in the same pleasing situation.

Letter to President Jefferson (1804)

But we must hold out to the end, without being weary of well doing, indulging no remission of labour while we are capable of any. Even a dying hand has sometimes done execution.

Dedication to Joshua Toulmin, D. D. in "Jesus and Socrates Compared. (1803)

That is right; I have now done

Last Words (1804)

Of Dr Priestley's theological works, he remarked, that they tended to unsettle every thing, and yet settled nothing.

Samuel Johnson (1770)

I am extremely concerned to hear that Dr. Priestley is so poorly provided for, while so many are rolling about here in gilt chariots, with very ample stipends. I admire his comprehensive genius, his perspicuity and vigor of composition, his indefatigable application, and his free, independent spirit, and wish it were in my power to do him any kind of service. It would give me great pleasure to see him well settled in America; though indeed I am inclined to think he can prosecute his learned labours to greater advantage in England. A man of his abilities would do honor to any of the colleges.

John Winthrop to Benjamin Franklin (1773)

... I rejoice to hear of your continual Progress in those useful Discoveries; I find that you have set all the Philosophers of Europe upon Fix'd Air; and it is with great Pleasure I observe how high you stand in their Opinion; for I enjoy my Friends' fame as my own.

The hint you gave me jocularly, that you did not quite despair of the Philosopher's Stone, draws from me a Request, that, when you have found it, you will take care to lose it again; for I believe in my conscience, that Mankind are wicked enough to continue slaughtering one another as long as they can find Money to pay Butchers.

Benjamin Franklin (1777)

... I always rejoice to hear of you being still employ'd in experimental Researches into Nature, and of the Success you meet with. The rapid Progress *true* Science now makes, occasions my regretting sometimes that I was born too soon. It is impossible to imagine the Height to which may be carried, in a thousand years, the Power of Man over Matter. We may perhaps learn to deprive large Masses of their Gravity, and give them absolute Levity, for the sake of easy Transport. Agriculture may diminish its Labour and double its Produce; all Diseases may by sure means be prevented or cured, not excepting even that of Old Age, and our Lives lengthened at pleasure even beyond the antediluvian Standard. O that moral Science were in as fair a way of Improvement, that Men would cease to be Wolves to one another, and that human Beings would at length learn what they now improperly call Humanity!

Benjamin Franklin (1780)

Chymistry was always an interesting pursuit with Dr. Johnson. Whilst he was in Wiltshire, he attended some experiments that were made by a physician at Salisbury, on the new kinds of air. In the course of the experiments frequent mention was made of Dr. Priestley. Dr. Johnson knit his brows, and in a stern manner inquired, "Why do we hear so much of Dr. Priestley?" He was properly answered, "Sir, because we are indebted to him for these important discoveries." On this Dr. Johnson appeared well content; and replied, "Well, well, I believe we are; and let every man have the honour he has merited.

James Boswell (1783)

Speaking of Dr. Pxxxxxxxx, (whose writing, I say, he estimated at a low rate), "You have proved him as deficient in probity as he is in learning. I called him an index scholar; but he was not willing to allow him a claim ever to that merit." He said, that "he borrowed from those who have been borrowers themselves, and did not know that the mistakes he had adopted had been quoted by others."

Mr Nichols as quoted by Boswell (1784)

To dispute with the good Dr. Priestley was the surest road to preferment. He had already *made two bishops* and there were several heads which wanted mitres.

Edmund Burke (1791)

The wild gas, the fixed air, is plainly broke loose: but we ought to suspend our judgment until the first effervescence is a little subsided, till the liquor is cleared, and until we see something deeper than the agitation of a troubled and frothy surface.

Edmund Burke (1791)

Remember me affectionately ... to the honest heretic Dr. Priestley. I do not call him honest by way of distinction, for I think all the heretics I have known have been virtuous men. They have the virtue of fortitude, or they would not venture to own their heresy; and they cannot afford to be deficient in any of the other virtues, as that would give advantage to their many enemies ... Do not however mistake me. It is not to my good friend's heresy that I impute his honesty. On the contrary 'tis his honesty that has brought upon him the character of heretic.

Benjamin Franklin to Benjamin Vaughn

... The name of Joseph Priestley will long be remembered among all enlightened people; and there is no doubt that England will one day regret her ungrateful treatment to this venerable and illustrious man. His persecutions in England have presented to him the American Republic as a safe and honourable retreat in his declining years; and his arrival in this City calls upon us to testify our respect and esteem for a man whose whole life has been devoted to the sacred duty of diffusing knowledge and happiness among nations ...

N. Y. Newspaper Editorial (1794)

I wish he had rather fixed his residence in Botany Bay as I think our people are mad enough already.

Charles Nisbit, Principal of Dickinson College (1794)

Lo! Priestley there, patriot, and saint, and sage,
Him, full of years, from his loved native land
Statesman blood-stained and priests idolatrous
By dark lies maddening the blind multitude
Drove with vain hate: Calm, pitying, he retired,
And mused expectant on these promised years.

Samuel Taylor Coleridge (1796)

I do not think it wise to execute the alien law against poor Priestley at present. He is as weak as water, as unstable as Reuben, or the wind. His influence is not an atom in the world.

President Adams to Secretary of State Pickering (1799)

Yours is one of the few lives precious to mankind, and for the continuence of which every thinking man is solicitous.

Thomas Jefferson to Joseph Priestley, March 21st (1801)

As the storm is now subsiding, and the horizon becoming serene, it is pleasant to consider the phenomenon with attention. We can no longer say there is nothing new under the sun. For this whole chapter in the history of man is new. The great experiment of our Republic is new. Its sparse habitation is new. The mighty wave of public opinion which has rolled over it is new. But the most pleasing novelty is its so quickly subsiding over such an extent of surface to its true level again. The order and good sense displayed in this recovery from delusion, and in the momentous crisis which lately arose, really bespeak a strength of character in our nation which augurs well for the duration of our Republic; and I am much better satisfied now of its stability than I was before it was tried.

Thomas Jefferson to Joseph Priestley, March 21st (1801)

Scarcely anything that has happened to me, in my curios life has made a deeper impression upon me, than that such a learned ingenious scientific and talented madcap as Cooper, could have had influence enough to make Priestley my enemy.

John Adams to Thomas Jefferson (1813)

Hitherto philosophy has been chiefly conversant about the more sensible properties of bodies; electricity, together with chymistry, and the doctrine of light and colours, seems to be giving us an inlet into their internal structure, on which their sensible properties depend. By pursuing this new light therefore, the bounds of natural science may possibly be extended, beyond what we can now form an idea of.

Autobiography

When I began these experiments, I knew very little of chemistry ... But I have often said that upon the whole this circumstance was no disadvantage to me; as in this situation I was led to devise an apparatus of my own, adapted to my peculiar views. Whereas, if I had been previously accustomed to the usual chemical processes, I should not have so easily thought of any other; and without new modes of operation I should hardly have discovered anything materially new.

JCR, p. 71

I own I had that expectation, when I first put a sprig of mint into a glass jar, standing inverted in a vessel of water; but when it had continued growing there for some months, I found that the air would neither extinguish a candle nor inconvenience a mouse.

I have had some appearances, which extraordinary as it will seem, make it rather probable, that light is necessary for the formation of this substance (algae).

I have discovered what I long have been in quest of, *viz.*, that process in nature by which air, made noxious by breathing, is restored to its former salubrious condition.

Three quotations in letter to Theophilis Lindsey (1771)

All the time that I was making these experiments, I wrote to my friends about them, particularly to Mr. Magellan, and desired him to communicate my observations to you, as well as to others; but I believe you had not heard of them; so that what you did with leaves was altogether independent of what I was doing with *whole plants*. The same summer, and the same sun, operated for us both, and you certainly published before me.

Letter to Jan Ingenhauz

As I never made the least secret of anything that I observe, I mentioned this experiment also, as well as those with m*ercurius calcinatus*, and the red precipitate, to all my philosophical acquaintances at Paris, and elsewhere; having no idea at the time, to what these remarkable facts would lead.

Autobiography

The feeling of it to my lungs was not sensibly different from that of common air; but I fancied that my breast felt peculiarly light and easy for some time afterwards. Who can tell but that, in time, this pure air may be fashionable article of luxury. Hitherto only two mice and myself have had the privilege of breathing it.

(1774)

From the greater strength and vivacity of the flame of a candle, in this pure air, it may be conjectured, that it might be peculiarly salutary to the lungs in certain morbid cases, when the common air would not be sufficient to carry off the phlogistic putrid effluvium fast enough. But, perhaps, we may also infer from these experiments, that though pure dephlogisticated air might be useful as a medicine, it might not be so proper for us in the usual healthy state of the body; for as a candle burns out much faster in dephlogisticated than in common air, so we might, as may be said, live out *too fast*, and the animal powers be too soon exhausted in this pure kind of air. A moralist, at least, may say, that the air which nature has provided for us is as good as we deserve.

Experiments and Observations on Different Kinds of Air (1776)

The more we see of the wonderful structure of the world, and of the laws of nature, the more clearly do we comprehend their admirable uses, to make all the percipient creation happy; a sentiment, which cannot fill the heart with unbounded love, gratitude and joy.

The History and Present State of Electricity (1776)

The greater is the circle of light, the greater is the boundary of the darkness by which it is confined. But, notwithstanding this, the more light we get, the more thankful we ought to be. For by this means we have the greater range for satisfactory contemplation.

Experiments and Observations on Different Kinds of Air (1790)

There have been few, if any, revolutions in science so great, so sudden, and so general, as the prevalence of what is now usually termed the *new system of chemistry*, or that of *Antiphlogistians*, over the doctrine of Stahl, which was at one time thought to have been the greatest discovery that had ever been made in science ... Though there had been some doubt of the existence of such a principle as that of *phlogiston*, nothing had been advanced that could have laid foundation of another system before the labours of Mr Lavoisier and his friends, from whom this new system is often called that of the French.

Considerations on the Doctrine of Phlogiston
and the Decomposition of Water (1796)

But I consider no hypothesis as of any use but to assist us in discovering new facts, which new facts serve in turn to correct the hypothesis; till, at length, by this method of approximating we shall have discovered all the facts, and have got a perfect theory.

Speculation is the cheapest of commodities.

In forming any general theory we must content ourselves with the fewest difficulties.

Though the title of this work expresses perfect confidence in the principles for which I contend, I still shall be ready to adopt those of my opponents, if it appears to me that they are able to support them. Nay the more satisfied I am with the doctrine of phlogiston, the more honourable I shall think it to give up on conviction of its fallacy.

Doctrine of Phlogiston Established (1801)

A philosopher who has long been attached to a favourite hypothesis, and especially if he has distinguished himself by his ingenuity in discovering or pursuing it, will not sometimes be convinced of its falsity by the plainest evidence of fact.

idem

... phlogiston, if there be such a thing as phlogiston.

idem

Priestley and History

In order to facilitate the advancement of all the branches of useful science, two things seem to be principally requisite. The first is an historical account of their rise, progress, and present state; and the second, an easy channel of communication for all new discoveries. Without the former of these helps, a person every way qualified for extending the bounds of science, labours under great disadvantages; wanting the lights which have been struck out by others, and perpetually running the risk of losing his labour, and finding himself anticipated in the discoveries he makes, which is a great mortification and discouragement.

The History and Present State of Discoveries relating to Vision, Light, and Colours

It will be seen, in the preface to the first edition of the History of Electricity, that I then considered the history of *all* the branches of experimental philosophy as too great an undertaking for any one person; but, like the fox with respect to the lion, a nearer view has familiarized it to me, and I now look upon it not only without dread, but with a great deal of pleasure; considering it not only as a very practicable *business*, but even as an agreeable *amusement*.

idem

I have adopted the *historical method*, because it appears to me to have many obvious advantages over any other for my purpose; being peculiarly calculated to engage the attention, and to communicate knowledge with the greatest ease, certainty, and pleasure. Moreover, the recital of the labours of philosophers in an historical method gives a writer a better opportunity than a systematical method would do, of transmitting them to posterity in such a manner as will operate most powerfully on the minds of the readers, ...

idem

These connections of man with man are every day growing more extensive. The most distant parts of the earth are now connected; every part is every day growing still more necessary to every other part; and the nearer advances we make to general happiness, and the more commodious over circumstances in this world are made for us, the more intimately and extensively we become connected with, and the more closely we are dependent upon one another.

JRC p. 78

We scruple not to plant trees for the benefit of posterity. Let us likewise now sow the seeds of truth for them ... I do not write this from a persuasion that everything I have myself contended for is indisputably true. On the contrary, I have for the sake of discussion, hazarded many things, and shall probably hazard many more; and I have actually changed my opinion, theological as well as philosophical, which I have advanced since I was a writer.

JRC, p. 80

Because common sense is a sufficient guard against errors in religion, it seems to be taken for granted, that common sense is sufficient instructor also [for children]; whereas, in fact, without positive instruction, men would naturally have been mere savages with respect to religion; as, without similar instruction, they would have been savages with respect to the arts of life and the sciences.

JRC, p. 67

There are many things in the system of nature, as tempest, lightening, diseases and death, which greatly terrify and annoy us, and which are often the occasion of much pain and distress; but these evils are only partial; and when the whole system, of which they are a part, and necessary consequences, is considered, it will be found to be, as far as we can judge, the best, and the most friendly upon the whole, and that no other general laws, which should obviate and exclude these evils, would have been productive of so much happiness.

JRC, p. 74

What Dr. Priestley added in discoursing from his written lectures (most of which are since published to the world) was pointedly and clearly illustrative of the subject before him, and expressed with great simplicity and distinctness of language, though he sometimes manifested that difficulty of utterance which he mentions in the Memoirs of his life. At the conclusion of his lecture he always encouraged the students to express their sentiments relative to the subjects of it, and to urge any objection to what he had delivered without reserve. It pleased him when any one commenced such a conversation.

Autobiography

When, for the sake of a little more reputation, men can keep brooding over a new fact, in the discovery of which they might, possibly, have very little real merit, till they think they can astound the world with a system as *complete* as it is *new*, and give mankind a high idea of their judgment and penetration; they are justly punished for their ingratitude to the foundation of mankind, in finding their boasted discoveries anticipated, and the field of honest fame pre-occupied, by men, who, from a natural ardour of mind engage in philosophical pursuits, and with an ingenious simplicity immediately communicate to others whatever occurs to them in their inquiries.

Experiments and Observations on Different Kinds of Air (1790)

I am sorry to have occasion to observe, that natural science is very little, if at all, the object of *education* in this country, in which many individuals have distinguished themselves so much by their application to it. And I would observe that, if we wish to lay a good foundation for philosophical taste, and philosophical pursuits, persons should be accustomed to the sight of experiments, and processes, in *early life*. They should, more especially, be early initiated in the theory and practice of *investigation*, by which many of the old discoveries may be made to be really *their own* ...

Experiments and Observations on Different Kinds of Air (1790)

At all events, however, the curiosity and surprize of young persons should be excited as soon as possible; nor should it be much regarded whether they properly understand what they see, or not. It is enough, at the first, if striking facts make an impression on the mind, and be remembered. We are, at all ages, but too much in haste to *understand*, as we think, the appearances that present themselves to us. If we could content ourselves with the bare knowledge of new *facts*, and suspend our judgment with respect to their causes, till, by their analogy, we were led to the discovery of more facts, of a similar nature, we should be in a much surer way to the attainment of real knowledge.

Experiments and Observations on Different Kinds of Air (1790)

Whatever be the qualification of your tutors, your improvement must chiefly depend upon yourselves. They cannot *think* or *labour* for you. They can only put you in the best way of thinking and labouring for yourselves. If, therefore, you get knowledge, you must acquire it by your own industry. You must form all conclusions, and all maxims, for yourselves, from premises and *data* collected, and considered by yourselves. And it is the great object of this institution to remove every bias and mind can lie under, and give the greatest scope to true freedom of thinking and inquiry.

Heads of Lectures (1794)

... A certain degree of vanity is, therefore, excusable in young persons; and, indeed, it is by means of it that they are excited to exert themselves in a manner that they would not otherwise have done. But be careful that this temper be not indulged to excess, for it will then be found to have serious ill consequences; the least of which is the precluding future improvement, from being already satisfied with yourselves, and conscious of a sufficient superiority over others.

Heads of Lectures (1794)

Also, as in old cities many buildings will fall to decay, while new ones are added; you must expect to forget much of what you now know. No man can give equal attention to every object; and as we advance in life, we, in general, only learn new things at the expense of the leading principles, which remain with us; while the more useless ones, things to which we give less attention because we find them to be of less use, disappear. Yet it is no uncommon thing for ingenious students to despise old scholars who are not so ready in the *minutiae* of literature, though they may have forgotten more than those youths ever knew, and may retain what they cannot acquire without forgetting as much.

Heads of Lectures (1794)

I think it most advisable not to trouble beginners with more than a large outline of any branch of science. By this means they are not fatigued by too long an attention to any one subject, a greater variety of articles may be brought before them, and in future life they may pursue any of them as much farther as their inclination may dispose, and their ability and opportunity shall enable, them to do it.

Preface to Heads of Lectures (1794)

In general, I would neither conceal from young persons the knowledge of vice, nor deny that temporal advantages and pleasures may attend vicious indulgences; but let them always be given to understand, that those advantages and pleasures are dearly bought.

JRC, p. 67

Priestley and Politics

The good and happiness of the members, that is of the majority of the members, of any state, is the great standard by which everything relating to that state must finally be determined.

"Essay on the First Principles of Government" (1768)

A spider in his natural size is only a spider ugly and loathsome, and his flimsy net is only fit for catching flies. But good God! Suppose a spider as large as an ox ... What would they do if they had power commensurate to their malice. God forbid I ever should have a despotic master – but if I must, my choice is made, I will have Louis XVI rather than Monsieur Brissot or Chabot; rather George III or IV than Dr. Priestley ... persons who would not load a tyrannous power by the poisoned talents of a vulgar low bred insolence.

Edmund Burke on Joseph Priestley (1792)

I the less wonder at this power of imagination and prejudice, and this stupefaction of your rational faculties in matters of religion, as it is apparent that you have been under a similar suspension of your reason, and equally under the power of imagination, in your views of the principles of civil government. Such, Sir, is "your proud submission, and the subordination of your very heart", to princes, and nobles; such devotion to rank and sex, in conjunction with your religious enthusiasm, that one might suspect that your book was composed after some solemn vigil, such as watching your arms at the shrine of the blessed virgin; after which you issued forth the champion, in form, of religion, of monarchy, and of the immaculate virtue of all handsome queens.

Joseph Priestley on Edmund Burke

It is with very sensible regret that I find Mr. Burke and myself on the two opposite sides of any important question, and especially that I must now no longer class him among the friends of what I deem to be the cause of liberty, civil or religious, after having, in a pleasing occasional intercourse of many years, considered him in this respectable light. In the course of his public life, he has been greatly befriended by the Dissenters, many of whom were enthusiastically attached to him; and we always imagined that he was one on whom we could depend, especially as he spoke in our favour in the business of subscription, and he made a common cause with us in zealously patronizing the liberty of America.

Joseph Priestley Letter to the Right Honorable Edmund Burke

I am proud of feeling myself a man among men, and I leave it to you, Sir, to be "proud of your obedience, and to keep alive" as well as you can 'in servitude itself the spirit of exalted freedom'. I think it much easier, at least, to be preserved out of a state of servitude than in it. You take much pains to gild your chains, but they are chains still.

-idem

The greatest instrument in the hand of the divine providence of this progress of the species towards perfections, is society and consequently government.

An Essay on the First Principles of Government (1771)

Priestley and Religion

The scheme of philosophical necessity has been shewn to imply a chain of causes and effects, established by infinite wisdom, and terminating in the greatest good of the whole universe; evils of all kinds natural and moral being admitted, as far as they contribute to that end, or may be in nature of things inseparable from it.

The Doctrine of Philosopical Necessity Illustrated (1777)

... the church of Rome, of which you are a member; but which I consider as properly *antichristian*, and a system of abominations little better than heathenism.

Letter to Abbé Boscovich (1778)

The morning is upon us, and we cannot doubt that the light will increase, and extend itself more and more unto the perfect day. Happy are those who contribute to diffuse the pure light of this everlasting gospel.

Joseph Priestley (1782)

Let us not, therefore, be discouraged, though for the present we should see no great number of churches professedly Unitarian ... We are, as it were, laying gunpowder, grain by grain, under the old building of error and superstition, which a single spark may hereafter inflame; in consequence of which that edifice, the erection of which has been the work of ages, may be overturned in a moment, and so effectually as that the same foundation can never be built on again ... And til things are properly ripe for such a revolution, it would be absurd to expect it, and in vain to attempt it.

(1787)

All I wish as a Christian from the powers of this world is that they would not intermeddle at all ... and that they would give no countenance whatever to any mode of it, my own or others, but show so much confidence in the principles of what they themselves deem to be true religion, as to think it able to guard itself.

I never preached a political sermon in my life, unless such as, I believe, all Dissenters usually preach on the 5th November [Guy Fawkes Day], in favour of civil and religious liberty, may be said to be political.

"Gunpowder Joe" Priestley JRC, p. 42

I have written with as much impartiality as I am capable of; following such lights as my reading and observation have been able to supply, I pretend not, however, to have written under the influence of all the virtues, that of discretion, in subjects of this nature, being almost out of my system.

JRC, p. 37

My Arian friends will think that I bear peculiarly hard upon them, and I frankly acknowledge it. I think theirs to be an hypothesis equally destitute of support in the scriptures, in reason, and in history.

JRC, p. 65

... [young ladies] seem to have no objection to a suitor on account of his illicit amours; imagining perhaps that a reformed rake will make the best husband ... Would this amiable part of our species only do themselves the justice to insist upon the same strict chastity and honour with respect to men, which men universally insist on with respect to them, our sex would, no doubt be as virtuous as theirs, and they would make much better husbands and fathers than they now do.

JRC, p. 58

Persons who have only one object of pursuit, never fail to over-rate it, and of course to undervalue things. I would farther observe, that the attention I have given to *theology* (which, by the way, is my original and proper province, and for which I may, therefore, be allowed to have a justifiable predilection) does not engross so much of my time as some persons may imagine.

Experiments and Observations on Different Kinds of Air (1790)

Neither the tenor of my theological writings in general, nor the manner in which I have written in defence of religious liberty in particular, were calculated to promise me the thanks of the bulk of Dissenters. But I think it now less honourable to be the hunted deer, shunned by his companions, than the leader of the peaceful herd.

Orthodoxy is my doxy; heterodoxy is another man's

Autobiography of Joseph Priestley

Selected References

1. Autobiography of Joseph Priestley, edited and with a long introduction by Jack Lindsay, Adams and Dart, Bath, England (1970). This is probably the best available source of Priestley's short, informative, but largely unrevealing autobiography.
2. *The Enlightenment of Joseph Priestley: A Study of His Life and Work from 1733–1773*, Robert E. Schofield, The Pennsylvania State University Press, University Park, PA (1997). This is the first volume of what is clearly to be the definitive biography by the ranking Priestley scholar, Robert Schofield. A second volume has been announced for the late summer of 2004. Indispensable.
3. A Scientific Autobiography of Joseph Priestley (1733–1804), Robert E. Schofield, The M. I. T. Press, Cambridge, MA, (1966). This section of those of Priestley's surviving letters that have significant scientific content (with extensive commentary by Schofield) provides by far the best introduction for the scientific reader.
4. *The Lunar Society of Birmingham*, Robert E. Schofield, Oxford University Press, (1963), Priestley is only one of the stellar list of characters who were associated with The Lunar Society, the intellectual heart of the early Industrial Revolution.
5. Motion Towards Perfection: The Achievement of Joseph Priestley, ed. A. Truman Schwartz and John G. McEvoy, Skinner House books, Boston, MA (1990). A very wide-ranging and readable collection of essays largely drawn from a commemorative symposia held in Washington and London in 1983. Particularly noteworthy is John Hedley Brooke's essay "A Sower Went Forth".
6. *Joseph Priestley: A Comet in the System*, John Ruskin Clark, The Friends of Joseph Priestley House, Inc., Northumberland, PA (1994). Clark is a Unitarian minister who has spent much of his life reading voluminous works of Joseph Priestley. The book is enlivened by numerous quotations, a number of which I have appropriated. Since Clark cites references from virtually inaccessible 26 volumes of Priestley's non-scientific works issues in 1817–1832. I simply cite "JRC" plus the page number.
7. *Joseph Priestley: Adventurer in Science and Champion of Truth*, F. W. Gibbs, Nelson, London (1965). A useful general introduction with emphasis on Priestley's scientific contributions.

Index

Authors of articles are in bold face

Excluded from the list are (i) Joseph Priestley (because of extensive appearance in the book), (ii) Author names appearing exclusively in reference lists, (iii) Character names in the Adam Strickson play, (iv) Instrument manufacturers, and (v) Biblical characters.